Traditional Knowledge for Green Innovation with Japanese Persimmon Tannin, Kakishibu

Kei Kano • Sujatha Raman • Dan Santos
Editors

Traditional Knowledge for Green Innovation with Japanese Persimmon Tannin, Kakishibu

Springer

Editors
Kei Kano
Faculty of Education
Shiga University
Otsu, Shiga, Japan

The Australian National Centre
for the Public Awareness of Science
Australian National University
Canberra, ACT, Australia

Sujatha Raman
The Australian National Centre
for the Public Awareness of Science
Australian National University
Canberra, ACT, Australia

Dan Santos
The Australian National Centre
for the Public Awareness of Science
Australian National University
Canberra, ACT, Australia

ISBN 978-981-95-7804-7 ISBN 978-981-95-7805-4 (eBook)
https://doi.org/10.1007/978-981-95-7805-4

The original submitted chapters 2, 3, 4, 5, 6, and 7 have been translated into English. The translation was done using artificial intelligence. A subsequent revision was performed by the author(s) and editors to further refine the work and to ensure that the translation is appropriate concerning content and scientific correctness. It may, however, read stylistically different from a conventional translation.

This Springer imprint is published by the registered company Springer Nature Singapore Pte Ltd.
The registered company address is: 152 Beach Road, #21-01/04 Gateway East, Singapore 189721, Singapore

Preface

Even I, a Japanese, have not been aware of Kakishibu for a long time. I was struck by its water-repellent capabilities and minimal environmental impact. Furthermore, I was impressed to learn that it is produced from the fermented juice of astringent persimmons. This was my genuine reaction upon initially encountering Kakishibu, a traditional Japanese product with a history spanning more than a millennium.

My encounter with Kakishibu dates back to 2019. Until then, I had been collaborating with a high school teacher, Toshihiko Matsuda, the author of Chap. 6, for many years in the development of STEM education programmes that focus on traditional Japanese technologies such as Kyoto vegetables and Japanese tea. When he was transferred to Kizu High School in Kizugawa City, Kyoto Prefecture, one of the three major Kakishibu production regions (Kyoto, Hiroshima, and Gifu) in 2018, he initiated business applications of Kakishibu with his high school students. I first encountered Kakishibu when high school students introduced their business plans and demonstrated a water-repellent experiment using a newspaper coated with Kakishibu.

He and his high school students requested that I develop a STEM education programme for children and guardians concerning Kakishibu. I promptly commenced the development of this programme with my undergraduate students, and the first workshop was conducted in January 2020. At that time, in contrast to the STEM education programme presented in Chap. 8, the focus of the programme was minimal regarding environmental issues and was primarily centred on water repellence.

Then, the COVID-19 pandemic struck, and we had no choice but to give up implementing the STEM education programme. As described in Chap. 6, the pandemic also became a major obstacle to the realisation of students' business plans to solve environmental problems and contribute to the UN's Sustainable Development Goals (SDGs) using Kakishibu. Nevertheless, we decided to do what we could even during the pandemic, and, as explained in Chap. 8, we developed and implemented a STEM education programme in which participants could participate online after receiving experimental equipment by mail. We also reorganised the content of the STEM education programme to include environmental issues and contributions to the SDGs.

One of the major turning points after that was my visit to the Australian National Centre for the Public Awareness of Science (CPAS) at the Australian National University (ANU) in August 2022, where we began collaborative research with Dr Graham Walker, co-author of Chap. 8, to add elements of Culturally Relevant Pedagogy and Utility Value Intervention to the STEM education programme. This led to a greater focus on the cultural and historical aspects of Kakishibu and traditional knowledge.

Another significant milestone was the seminar I delivered at CPAS, ANU, in March 2023, during which attendees had the opportunity to engage with a STEM education programme centred on Kakishibu. Professor Sujatha Raman, who is the co-editor of this publication and the author of Chaps. 1 and 9, along with her doctoral student Bangle Wu, co-author of Chap. 9, both of whom were present in the audience, facilitated a deeper discussion contrasting Kakishibu with plastics and Per- and Polyfluoroalkyl Substances (PFAS). They have been leading the UNESCO Chair in Science Communication for the Public Good since 2022. The UNESCO Chair covered the topic of green innovation and the plastic society, as described in Chap. 1.

This discourse led to subsequent interviews with stakeholders involved in Kakishibu and contributed to the creation of a novel green innovation framework based on these interviews. Moreover, this discussion heightened my awareness of the importance of integrating the Japanese views of nature with Western scientific approaches, which underpin both Kakishibu and plastics. This conceptual discussion is elaborated in Chap. 2.

Viewing it from this perspective, it becomes evident once again that the primary entities within the narrative of Kakishibu are the *shibuya*, traditional small-scale Kakishibu manufacturing and sales companies. Chapter 3 documents the history and development of Kakishibu authored by a shibuya, Mimasu Kashichi Shouten Co., Ltd., which was founded in 1872. Readers will observe how the company has continually embraced new challenges while honouring its traditions. There is no doubt that *shibuya* remains at the core of Kakishibu; however, this publication also explores the pathways undertaken by scientists and engineers in promoting the integration of Western scientific principles with traditional practices, thereby contributing to green innovation (Chaps. 4, 5, and 7). These discussions extend beyond mere proposals for enhancing Kakishibu through scientific and technological means to encompass considerations on the responsible utilisation of such technologies and the necessary modifications in our actions and lifestyles.

As previously outlined, this publication employs Kakishibu as a case study to examine the philosophy, science, technology, and responsibilities inherent in traditional Japanese knowledge. It also delineates a pathway towards sustainable innovation by integrating it with Western scientific principles. While the book is organised to be read sequentially from Chap. 1, readers may opt to commence with the chapters that most engage their interest.

Finally, I would like to make a request to readers who are interested in Kakishibu and the green innovation that utilises it. That is to emphasise the utmost respect for traditional knowledge. As stated in Chap. 9, rapid and large-scale development is not advisable, and an appropriate use, design and solutions must be maintained. Specifically, one potential mechanism is the fair trade of Kakishibu, including the protection of traditional production techniques. It would be beneficial to keep Joseph Needham's words, inspired by Taoism, 'while it is impossible to avoid acquiring knowledge, we must endeavour not to employ knowledge that poses a threat to humanity' (Nakayama et al., 1988), firmly in mind, and to ensure that Kakishibu is not exploited and modified too much.

Reference

Nakayama, S., Matsumoto, S., & Ushiyama, K. (1988). *Josefu Nidamu no sekai (The world of Joseph Needham)*. Nihon Chiiki Shakai Kenkyujo.

Otsu, Shiga, Japan Kei Kano

Acknowledgements

This book would not have been possible without the support of the lab members (Kaneshiro, R., Nakashima, W., Takatsuka, M., Morisaki, R., Arima D., Hagihara, K., Tsushi, Y., Sano R., and Azai, H.) at Shiga University and the Australian National Centre for the Public Awareness of Science (CPAS), Australian National University (ANU) support for the UNESCO Chair in Science Communication for the Public Good. We also thank Yasunori Yamashita and Yuki Satsuma for participating in the interviews, Questacon members for their advice on the STEM workshop about Kakishibu, and Mei Han Lee and Momoko Asawa for their valuable editorial assistance. This work was partly supported by JSPS KAKENHI Grant Number 21KK0230, and this book was published with the support of the Shiga University Press Grant Programme.

Contents

Contributors

Tomokazu Goto Mimasu Kashichi Shouten Co., Ltd., Kyoto, Japan

Toshiharu Goto New Business Management Division, Maxell, Ltd., Kyoto, Japan

Kei Kano Faculty of Education, Shiga University, Otsu, Shiga, Japan

The Australian National Centre for the Public Awareness of Science, Australian National University, Canberra, ACT, Australia

Toshihiko Matsuda Ikubunkan Institute of Education, Bunkyo City, Tokyo, Japan

Tomoaki Matsuo The Scientific Research Association for Kakishibu and Kakitannin, Kyoto, Japan

Takeo Mimasu Mimasu Kashichi Shouten Co., Ltd., Kyoto, Japan

Yoshitsugu Mimasu Mimasu Kashichi Shouten Co., Ltd., Kyoto, Japan

Sujatha Raman The Australian National Centre for the Public Awareness of Science, Australian National University, Canberra, ACT, Australia

Dan Santos The Australian National Centre for the Public Awareness of Science, Australian National University, Canberra, ACT, Australia

Shota Taniguchi New Business Management Division, Maxell, Ltd., Kyoto, Japan

Graham Walker The Australian National Centre for the Public Awareness of Science, Australian National University, Canberra, ACT, Australia

Bangle Wu The Australian National Centre for the Public Awareness of Science, Australian National University, Canberra, ACT, Australia

Yuka Yokoyama New Business Management Division, Maxell, Ltd., Kyoto, Japan

Part I
Introduction and Overview: Putting Japanese Views of Nature in the Global Context of Scholarly and Public Debates on Knowledge-Systems

Chapter 1
Green Innovation and the Plastic Society

Sujatha Raman and Dan Santos

Abstract This chapter introduces the volume, outlining how it came about, the problem it is responding to, and the approach taken in individual chapters. In doing so, it situates the history and future of Kakishibu in the context of major concerns surrounding the proliferation and endurance of petrochemical plastics, and debates in the social sciences about the appropriate way to intervene in such crises through innovation. The opportunity for collaboration between the Kakishibu community of experts in Japan and a team of social science researchers from Japan and Australia is described, with the common focal point being a shared interest in new ways of doing and researching science communication. This chapter distinguishes between three ways of approaching innovation in response to the planetary crisis posed by the plastic society: technology led, socio-technical, and alternative innovation pathways. While Kakishibu offers a promising way forward, innovating with this traditional material will require grappling with social and cultural norms across industrial systems, not technological visions alone.

Keywords Planetary crises · Petrochemical plastics · Green innovation · Socio-technical · Innovation pathways · Sustainability transitions · Technological breakthroughs · Kakishibu

1.1 Introduction

This book aims to stimulate ways of thinking and promoting green innovation beyond business-as-usual. Future historians may come to see 2025, the year in which this book was finalised, as a turning point in the global order when conventions fell apart and chaos reigned at many levels. Yet, two themes remain stubbornly

S. Raman (✉) · D. Santos
The Australian National Centre for the Public Awareness of Science,
Australian National University, Canberra, ACT, Australia
e-mail: Sujatha.Raman@anu.edu.au; Dan.Santos@anu.edu.au

K. Kano et al. (eds.), *Traditional Knowledge for Green Innovation with Japanese Persimmon Tannin, Kakishibu*, https://doi.org/10.1007/978-981-95-7805-4_1

resilient through the uncertainty and upheaval. One is the imperative posed by planet-scale environmental crises which people have been grappling with in everyday life as well as through their roles in civic and cultural movements, governments, trans-governmental institutions, and their professions. The second is the enduring fascination with technological innovation, valued both in its own right and as a way out of our planet-scale dilemmas. On the first theme of planetary threat, there is considerable political disagreement within and between countries. But on the second theme of innovation centred on novel technologies—or more accurately, the promise of future breakthroughs—there is far more consensus or, perhaps, a blind faith. This book is situated at the nexus of these enduring themes of hazard, on the one hand, and innovation, on the other.

We take the concern about planetary crises seriously, focusing in this instance on the specific threat posed by the proliferation of plastics. We also recognise the imperative to innovate, but here we depart from the faith in future technological breakthroughs as a straightforward solution to planetary crises, focusing instead on how we might re-imagine green innovation in ways that could be described as attentive to socio-technical, cultural, and ethical matters.

The UN Environment Programme (UNEP) brings together key dimensions of our contemporary challenges under the label of the 'triple planetary crisis'. The concept signals the interconnections between the problems posed by climate change, biodiversity loss, and environmental pollution. Although the trans-governmental consensus on their severity and on the shape of appropriate action is fragile, this does not change the fact that the problems are still staring us in the face. A critical aspect is what *The Lancet* describes as the 'grave, growing and under-recognised danger to human and planetary health' posed by the ubiquity and growth of petrochemical-based plastics (Landrigan et al., 2025). The fact that plastics are polluting the environment is now widely publicised, especially in light of growing attention to the impacts of microplastics on marine and freshwater systems (Frias & Nash, 2019). Plastics also exemplify the wider triple planetary crisis since their production is intertwined with habitat loss and greenhouse gas emissions as well as pollution (Bachmann et al., 2023).

Some responses to the plastics crisis revolve around political action at national and transnational levels with communities continuing to advocate for transformative change, including significant reductions in the production of petrochemical plastics, and mobilising of coalitions to counter powerful interests opposed to such change. Other work will need to be at the meso level, building systems to support manufacturers, businesses, and consumers weaning themselves off entrenched practices of plastic use and disposal in goods and in everyday life. A third and related response centres on devising alternatives to petrochemical-based plastics, which account for 98% of plastics produced today. Our book explores this third line of possibility focused on innovation. We do not assume that a single technological breakthrough will solve the plastics crisis; rather, we seek to open up consideration of diverse innovation pathways, and the values, cultures, and institutions that could support ways of meeting human needs without relying excessively on the plastics of today.

The focus of our inquiry is the fascinating history of Kakishibu, the Japanese term for fermented persimmon juice, which has been widely used since the pre--industrial age for making everyday items water-resistant, strong, and insect-proof. This book features members of a small and highly dedicated Japanese community working to recover and share traditional knowledge about Kakishibu's properties, and to stimulate appropriate ways of adapting this knowledge in response to the plastic waste crisis. Knowledge of and experience with Kakishibu represent an opportunity for green innovation, the nature of which we aim to unpack in the volume.

The term 'green innovation' tends to be associated with studies of business and management where it is defined in terms of organisational efforts to develop and adopt technologies that reduce pollution, carbon emissions, and related environmental risks (Takalo et al., 2021). At face value, this seems unobjectionable and, indeed, desirable. However, the language and practice of green innovation and green transformation, more broadly, has come to be problematised for reasons of what gets included or excluded. Social scientists in science and technology studies (STS) and development studies argue that we need to seriously examine: what gets defined as 'green', potentially to the exclusion of other worthy possibilities; who decides what the goals are and how to achieve them; and what sorts of expertise and technologies are included in these conversations and practices (Leach, 2015; Stirling, 2011).

Such issues are inherently value-based which means that technological solutions to environmental problems often raise ethical questions about who benefits and who might be negatively impacted. They also involve cultural differences between various settings, calling for efforts to illuminate alternative imaginaries of desirable futures and ways to get there (Beck et al., 2021). Finally, technological change, even forms considered to be desirable or in the public good, is never solely about tools, artefacts, or even economics. STS scholars describe innovation as socio-technical because there are issues to consider about how to assemble social norms and institutional practices of production and use, in conjunction with technical knowledge and practice, in order to accomplish change (Irwin, 2023). This chapter aims to position the story and possible futures for Kakishibu in the context of these social science insights and debates.

A famous STS history of Bakelite, commonly known as the first modern plastic (Knight, 2014), begins by observing 'The history of human use of plastic materials is as long as the history of humanity itself' (Bijker, 1987, p. 175). Bijker offers examples from ancient Egypt (resins) and Greece (amber). Popular histories tell the story of shellac in ancient India and China, and its subsequent travel to Europe in the 1400s (Fone, 2023), and the British chemist Charles Macintosh's development of water-resistant materials from naptha and rubber in the early 1800s (Knight, 2014; Fone, 2023). But what kinds of materials did people devise in other parts of the world to meet their needs, and how can these histories be of relevance today? The story of Kakishibu from Japan is a compelling one in this regard. In recovering its history and spotlighting the contemporary Kakishibu community, we hope to open up a pathway for a future in which the burden of the already existing plastic age must still be confronted, but where we countenance a broader canvas of possible innovations in the development and adoption of useful materials.

First, we briefly examine the contours of the plastics crisis in the next section before giving an account of how this book came about and the approach it takes. Next, we build on social science literature to provide an overview of different ways to conceptualise and orient innovation in response to this crisis. We then conclude with a few observations on the relevance of these themes for envisioning potential futures for Kakishibu as an innovative alternative to a future that is still dominated by petrochemical plastics.

1.2 Plastics and Pollution: Contours of a Pervasive Problem

Plastics have become ubiquitous in many of the products and goods that populate contemporary advanced industrial and industrialising societies. Several reasons account for the rapid proliferation of plastic, including its material properties (lightness, durability, versatility) and economic production costs (being a by-product of fossil fuel processing for energy). This has led to the widespread and pervasive presence of plastics in practically every domain of social life. However, it is now evident that plastics constitute a serious and significant threat to humans and nature (Landrigan et al., 2025; Macleod et al., 2021). Initially, their durability and long-life was viewed favourably, but now such qualities have become problematic features.

The increasing volume of plastics disposed as waste, and the limited availability of commensurate solutions to breakdown, recycle, and remove plastics, has meant that they remain and persist in environments for extended time periods. This protracted presence poses problems at several scales. At local or regional levels, plastics are present in a range of terrestrial and marine ecosystems with serious negative and lasting impacts on wildlife, soils, nutrient cycles, and human health (Li et al., 2021). Plastics are also a global problem, not only because they constitute 'novel entities' for which planetary limits need to be discerned and imposed (Steffen et al., 2015), but more broadly because they will likely constitute a key factor in future geological shifts to new epochs, such as the Anthropocene (Zalasiewicz et al., 2016). Archaeologists similarly argue that the ubiquity, resilience, and toxicity of plastics must be considered across space and deep time which, for them, means attending to how the plastic age is intertwined with extractivism, consumerism, climate change, and institutionalised patterns of waste production (Schofield et al., 2025).

The multi-scalar problems posed by plastics have led to an interest in similarly multi-scalar solutions. At the micro-level, there are efforts to understand purchasing, use and disposal practices, in order to intervene in and instigate changes to individual behaviours and household dynamics (e.g. Reynolds, 2025). Citizen science initiatives have contributed to reducing plastics in local environments through coordinated clean-up and recycling activities (Nelms et al., 2022), and activist groups have undertaken efforts to engage with businesses and governments to bring about impactful change (Dauvergne, 2023). Individual countries have also developed national regulations targeting plastic production, consumption, and disposal (Knoblauch & Mederake, 2021). Finally, at the global level, multi-lateral initiatives attempt to enforce or impose more consistent approaches to governing plastics. These include the UNEP Plastics Initiative (UNEP, 2025) to 'end global plastic pollution' across the full life cycle from production to use to disposal. This aspiration dates back to a commitment made in 2022, signed by 175 UN member states, to develop an international legally binding agreement to end plastic pollution, with ongoing meetings and discussions taking place to negotiate the specific details and terms of the agreement (Global Plastic Laws, 2025).

However, at the time of writing, negotiations between UN member-states on this ambitious Global Plastics Treaty concluded without an agreement (Stallard & Poynting, 2025). Unsurprisingly, the goal of curbing the production of oil-derived plastic—as opposed to focusing on improving recycling rates—remains a major sticking point. Social scientists also argue that a successful Treaty will require greater accountability, transparency, and meaningful inclusion of Indigenous rights--holders and civil society (Dauvergne et al., 2025).

But what role can, and should, science and technological innovation play in addressing plastics' problems? The unquestionable successes in rendering plastic a versatile, effective material are no doubt due, in large part, to scientific and technological ingenuity, both with respect to production and further upstream with respect to mineral extraction in the fossil fuel industry. As such, science and technological innovation bear some responsibility, and should play a significant role in developing commensurate solutions to addressing a problem they helped generate. Science- and technology-led innovations to address the plastics crisis are being investigated and deployed across research programs, business, and local government initiatives. There is emerging interest in research and development for novel materials which could represent more environmentally friendly options in place of standard plastics.

For example, one line of work focuses on technologies that collect plastics and prevent them from entering certain environments, such as oceans (Schmaltz et al., 2020). There is also a significant body of work on so-called bioplastics (Rujnić-Sokele & Pilipović, 2017); however, this is a label which has attracted scientific scrutiny for its ambiguous and sometimes misleading uses (Rosenboom et al., 2022). Bioplastics can refer to polymers made from renewable resources, but this does not guarantee the end product is biodegradable within specified environments and timelines (Folino et al., 2023; Rosenboom et al., 2022). Depending on their composition, materials labelled 'bioplastics' may also produce toxic materials upon decomposition. Furthermore, using renewable resources can have their own sustainability challenges arising from the nature, sourcing, and processing of biomass feedstocks, as previous work on bio-based liquid fuels has shown (Mohr & Raman, 2013).

In sum, new bio-based technological alternatives to plastics have their own limitations and shortcomings, and these will not be sufficient to address the growing problem and its complex dynamics. Hence, the re-consideration of traditional materials—such as Kakishibu—and appropriate options for adapting those to modern societies is particularly apt.

1.3 Context and Approach of the Book

This volume grew out of a collaboration between the Australian National University (ANU) and Shiga University aimed at exploring the promise and potential of Kakishibu as a socio-technical and cultural material which could represent a green innovation option in response to the planetary threat of plastic pollution. During one of his spells as a Visiting Fellow at the ANU Centre for the Public Awareness of Science (CPAS), the first editor, Kei Kano, ran a hands-on science communication workshop demonstrating the unique properties of Kakishibu. One of the themes of the ensuing discussion centred on prospects for increasing awareness outside of Japan of this traditional technology, perhaps with an eye towards promoting wider take-up of Kakishibu-based goods as an alternative to conventional fossil fuel-based plastics. The second editor, Sujatha Raman, was intrigued. She saw an opportunity to explore this question from the standpoint of research on diversifying innovation pathways, or what Stirling (2011) describes as 'pluralising progress'. Raman was leading the UNESCO Chair in Science Communication for the Public Good which seeks to build on research and practice in STS and reflexive science communication (Ishmael-Perkins et al., 2023) to explore appropriate ways of bringing science and innovation to bear on the Sustainable Development Goals (SDGs).

Research on diversifying innovation imaginaries and pathways aims to get us thinking more deeply about how innovation might be assembled and structured in favour of particular normative visions of the future, while foreclosing others (Beck et al., 2021; Stirling, 2011). Instead of taking for granted that every instance of technical ingenuity must be 'scaled up' (Pfotenhauer et al., 2022) in order to fit into the dominant socio-economic order, we are prompted to consider potential alternatives to that order. Therefore, rather than taking for granted that a solution to the plastics crisis ought to provide a one-on-one replacement for non-degradable plastics and leave the cultural and technical norms of a plastic society alone, we might instead attend to the possibility of transforming those norms.

Such alternatives become important because of the uncertainties and contradictions of deploying 'green' technology in a system organised to maximise the production and use of resources. These contradictions are especially apparent when traditional knowledge and technical expertise are extracted out of their original small-scale setting to fit into the needs of mass industrial societies. For example, the plant *Jatropha curcas* traditionally grows on degraded or marginal land and has been used as a source of fuel for rural communities in India. But when it came to be heralded as an ecological alternative to petroleum-based liquid fuel and cultivated in the form of large-scale plantations, *Jatropha*'s original properties no longer held true (Ariza-Montobbio & Lele, 2010). In this mass form, it disrupted traditional livelihoods, resource use, and biodiversity patterns, and failed to generate the kind of fuel yields expected from large-scale investments (Jewitt & Raman, 2017).

Out of these conversations was born our joint ANU-Shiga University research project, '*Innovation Frames for Social and Environmental Good: Envisioning Alternative Futures for Kakishibu Technology*'. We agreed that it would be interesting to find out how members of the Kakishibu community in Japan themselves thought about the future of this traditional technology. How did they make sense of the history, purpose, and practices of producing and using Kakishibu; and of the material and social infrastructures needed to support the growth of green innovation centred on Kakishibu's properties? At what scale or scales (local, regional, global) should this innovation be pursued? What might be some different future pathways for Kakishibu to take, and what social and environmental purpose would they be centred on?

To explore these questions, Kano and Raman, in collaboration with Bernice Wu (doctoral student at the time in CPAS, ANU), designed and conducted a piece of qualitative interview-based research in late 2023 with key members of the Kakishibu community in Japan. Dan Santos subsequently joined the UNESCO Chair team and brought his expertise in STS and environmental transitions to the editorial team for this volume.

Before we position the chapters that follow in terms of our conceptual interests, we want to draw the reader's attention to how we have adapted the qualitative research approach in this volume. Unlike standard qualitative work, we have sought to feature this research as a collaborative enterprise with Kakishibu community members. Their perspectives, expertise, and experience feature in their own voices in subsequent chapters rather than as mediated by us as editors. However, we use these introductory (Chaps. 1 and 2) and concluding (Chap. 9) chapters to position Kakishibu in terms of the broader themes of green innovation pathways that sparked this collaboration. Chapter 8 was written separately, featuring interests shared between Kei Kano and Graham Walker (CPAS, ANU) in hands-on science communication engagement activities with young people.

If more holistic and systemic approaches to addressing the pervasive problems posed by plastics are to be developed, we need to consider the role of innovation from a more integrative socio-technical and socio-ecological perspective. In the following section, we provide an overview of theoretical perspectives that position science as contributing in different ways to developing innovative solutions to environmental problems. This overview will help map out various options for how science may be better positioned to contribute to sustainable pathways towards using, managing, and living with (more green) plastics as well as without (petrochemical) plastic.

Scholars investigating prospects for transforming industrial societies in environmentally sustainable directions conceptualise this system-wide challenge as a 'socio-technical' one (Geels, 2004; Köhler et al., 2019). The concept of the socio-technical or, for some authors, 'socio-technological' (Miller, 2023), captures the idea that technologies necessarily embed and embody particular human values, relationships, and institutional orders. In other words, they are never merely embodiments of scientific and technical knowledge of nature. This means that transforming the technological orders which we know degrade nature—and in turn human and social relationships and interdependence with nature—cannot be a job for technical expertise alone. Rather, transformation calls for mobilising diverse forms of expertise, devising new models of valuing nature and human endeavour, and ultimately rethinking novelty or what is commonly valorised as novel in modern industrialised and industrialising societies inflected by 'Western' logics.

1.4 Conceptualising Green Innovation: Three Strands

Growing societal concerns with environmental problems have instigated an ongoing effort and debate around how to address these problems through sustainability initiatives. Innovation is often envisaged to play a central role in transformations towards more sustainable futures, especially innovation that is explicitly environmentally oriented, or 'green'. Nevertheless, there may be strong disagreements about what constitutes 'green innovation' and what sorts of transformation may be brought about, and are desirable through, such innovations. As social scientists have shown, such disagreements may revolve around different meanings around 'green' (Leach, 2015), along with different meanings around 'innovation', in both technical and normative terms. As such, it is important to differentiate between various approaches to conceptualising green innovation in order to appreciate their key assumptions and expectations. In this section, we group these approaches into three strands in order to draw attention to broad differences between them. This section is not intended to be comprehensive, but instead to paint a general illustration of the differences between innovation framings.

1.4.1 Technology-Led Approach

The dominant framing with respect to the role of innovation to address environmental issues has tended to centre technology as the locus for creating desirable, lasting change. This framing sees technological innovation, and the goods and products which are developed through such innovation, as the key driving force for instigating broader scale transformation towards more sustainable futures. In the business studies field, the term 'green innovation' is widely deployed to mean an organisational process that contributes to the creation of new technologies with the aim of reducing environmental risks, such as pollution and the harmful consequences of resource exploitation (Castellacci & Lie, 2017; Takalo et al., 2021).

This framing downplays or overlooks the role of other factors in shaping transformation possibilities, such as cultural norms and practices, within organisations but also beyond. As such, this framing may be described as 'technologically determinist', in that it upholds the belief that 'technological progress equals social progress' (Wyatt, 2007, p.169). Technological innovation, then, provides viable and necessary possible solutions to address pressing environmental problems. In their early days, the influential frameworks of ecological modernisation and sustainability transitions tended to focus on environmentally friendly technological change without looking into some of the normative divergences and alternatives that start to emerge in later work.

Ecological modernisation emerged in the 1980s as a conceptual approach for understanding environmental reform efforts in advanced industrial societies, especially in Western Europe (Mol et al., 2020). Early on, the concept prioritised 'a more pragmatic posture that created space for dialogue and negotiation between professionalised environmental movements, expanding and diversifying environmental states, and increasingly engaged private sector actors' (p. 7). As such, innovation within ecological modernisation is often commensurate with dominant industry perspectives within contemporary capitalist societies, privileging technology-focussed and market-oriented innovation pathways.

Early literature on sustainability transitions comprises another prominent approach to conceptualising environmental innovation and broader societal transformation. There are several theoretical models and frameworks within this literature (Geels, 2002). One popular model is the multi-level perspective, which is comprised of three levels. Niches are one level, where novel or radical innovations are developed. These innovations are largely protected from the existing market dynamics and status quo norms, regulations, and institutions found in another level: regimes. Broader scale events and dynamics at the landscape level then shape how niches interact with regimes to potentially galvanise transformation through the adoption and impact of niche-innovations. Nevertheless, despite emphasising the reciprocal interactions between levels, the key vehicle and driver of change remains technological innovation, at least in the original incarnations of this line of research. Technological innovation becomes the vehicle for instigating broader social and environmental transformations compared to the status quo.

In recent years, these social science traditions have grown and evolved considerably. For example, ecological modernisation thought has expanded to encompass more global perspectives, and prioritised the importance of understanding industry and consumer practices contributing towards environmental and sustainability goals. Taking stock of how European societies have in fact evolved in recent decades, one of the founders of this framework reflects on the fact that environmental improvements have been selective, often co-existing with polluting systems rather than transitioning out of them altogether (Jänicke & Jörgens, 2023). Similarly, recent work on sustainability transitions emphasises concerns around inclusivity and diversity in experiments in this domain, alongside broader ideas of social learning amidst uncertainty (Sengers et al., 2019). Köhler et al. (2019) observe that much remains to be done to put sustainability transitions work in conversation with emerging debates on limits to growth and innovations in alternative community economies. These developments underline the limits of a technology-centric approach to green innovation and begin to open up broader issues that need to be part of the picture.

1.4.2 Socio-technical Approach

The dominant framing of innovation that focusses on producing and disseminating technological artefacts has been subject to several lines of critique. One significant critique has foregrounded the need to consider the range of factors that inextricably shape technological innovation. In doing so, assumptions or expectations that technological innovations constitute, on their own or even primarily, 'fixes' to much more complex problems are problematised and contested.

Early versions of this critique were developed within Science and Technology Studies (STS) through arguments around the social shaping or construction of technology (e.g. Williams & Edge, 1996; Pinch and Bijker, 1984). These critiques draw upon examples to illustrate how decisions and trajectories for technological innovation are always more open-ended and contingent, and a range of social, political, and economic factors shape the range of pathways that could be taken. Therefore, the success and effectiveness of innovations require considering how, 'alongside narrowly "technical" considerations, a range of "social" factors affect which options are selected—thus influencing the content of technologies, and their social implications' (p.866). Technological determinism is instead replaced by technological contingency.

These critiques extend to the concept of a 'technological fix' (Rodner, 2004), in which technological innovations are designed specifically to address particular social or environmental problems. This work has not only revealed the shortcomings of investing too much hope, or hubris, in technological innovation to resolve complex and dynamic issues, but also demonstrated how additional unexpected problems may arise as well.

Levidow and Raman (2020) provide an illustrative example to critique ecological modernist imaginaries. Through examining imaginaries associated with low-carbon waste-energy policies in the UK, they demonstrate that dominant imaginaries in this domain revolve around prioritising technological innovation and market processes to address problems and pursue progress, and that the legitimacy of these imaginaries is enabled through broader policy and institutional arrangements. Furthermore, alternative imaginaries, which in this case revolved around eco-localisation, highlight how technological innovation necessarily involves social dynamics. This aligns with influential work in the sustainability transitions field, calling explicitly for a move away from new technologies to highlight the importance of social and economic system change (Schot & Steinmueller, 2018).

The critiques outlined in this strand are necessary, but not sufficient, for charting more constructive future pathways to address environmental problems through innovation. Some of the prior critical analysis is now leading to new lines of inquiry centred around reimagining innovation itself and opening up alternative innovation trajectories (Irwin, 2023; Joly, 2019). There is a new desire to build on critiques of the narrow scope of dominant imaginaries of how innovation is expected to unfold and for what purpose, and to consider the question of how to support better forms of innovation.

1.4.3 Envisioning Alternative Innovation Pathways

In adopting a more constructivist sensibility to understanding technological development, multiple roles and pathways for innovation become discernible, and more possible futures are opened up for scrutiny and social discussion. In doing so, several different innovation options may be pursued simultaneously, such that a one-size-fits-all approach, inattentive to context, will not be enough.

Scholars in STS have critiqued the generic celebration of 'innovation' because it is often equated with large-scale technological breakthroughs, irrespective of their biases, exclusions or, indeed, mismatch with societal priorities (Inverizzi et al., 2023). Breakthroughs may happen in controlled conditions, but getting them to work in real-world settings may be altogether different, involving challenges and impacts not previously envisaged (Glover, 2022). Technologies can and do solve many problems—to cite just one example, flue gas desulphurisation systems removed acid-rain-causing sulphur dioxide emissions from coal-fired gas plants. But the point is that many of our complex crises are less amenable to resolution with these sorts of technological fixes (Sarewitz & Nelson, 2008). The instinct to be wary of dominant ways of imagining innovative solutions to our collective problems extends equally to how problems are framed in the first place. Masco (2017) argues that in the twenty-first century, the language of crisis as a representation of collective problems has paradoxically allowed existing social and economic orders to be stabilised. Policymakers are able to appeal to technological fixes—which do not actually tackle the root causes of planetary challenges—instead of attending to the necessary task of transforming these orders.

These lines of analysis have led to an emerging interest in more system-based change in which transformations in sustainable directions point towards cultural, economic, and social system change in conjunction with the technologies appropriate to these purposes. Stirling (2011) prompts analysts to seriously grapple with the meanings and possibilities of diversity in technological innovation and change, in order to pluralise conceptions of progress beyond the technologically determinist ones that dominate contemporary imaginaries of innovation. Recognising and working through disciplinary diversity in understandings of innovation and transformation should occur alongside considerations of normative diversity, in order to attend to the full range of innovation pathways that could be undertaken.

Several scholars have given examples showing why it is viable to take seriously the hope for devising alternative socio-technical configurations supportive of, and enabled through, innovation (e.g. Santos, 2023). In examining the role of infrastructures in everyday life, Shove et al. (2018) affirm the '*variety* of infrastructure–practice relations' (p. 3; emphasis in original) through which norms and practices of consumption may be aligned and assembled, including in less resource-intensive ways. In another example, Shortall et al. (2015) analyse biorefineries as infrastructures capable of producing a range of critical resources. They describe not just the dominant imaginary (which focusses on biotechnological innovation), but also outline two alternative frameworks through which other socio-technical arrangements revolving less around biotechnological innovation are offered. They conclude that 'it is important to open up embedded assumptions within dominant visions of a technology and make space for alternatives' (p. 367). Opening up assumptions, expectations, and imaginaries is a critical step towards surveying a wider array of innovation pathways, but these efforts need to be provided with the latitude and support in policy and research practice to be considered viable and legitimate possibilities.

1.5 Conclusion

The following set of chapters capture different ingredients that will be necessary to consider to imagine and chart innovation futures with Kakishibu at the centre. They range from consideration of traditional Japanese ways of knowing to the deep expertise and work of core members of the Kakishibu community (including the history of this material and scientific and cultural properties, factors shaping its manufacturing and transformation into an industrial product, and the role of education and youth engagement). We then return to the theme of alternative green innovation pathways, distilling the insights gathered from the main chapters and exploring their implications in the final chapter of the volume.

References

Ariza-Montobbio, P., & Lele, S. (2010). Jatropha plantations for biodiesel in Tamil Nadu, India: Viability, livelihood trade-offs, and latent conflict. *Ecological Economics, 70*(2), 189–195. https://doi.org/10.1016/j.ecolecon.2010.05.011

Bachmann, M., Zibunas, C., Hartmann, J., Tulus, V., Suh, S., Guillén-gosálbez, G., & Bardow, A. (2023). Towards circular plastics within planetary boundaries. *Nature Sustainability, 6*, 599–610. https://doi.org/10.1038/s41893-022-01054-9

Beck, S., Jasanoff, S., Stirling, A., & Polzin, C. (2021). The governance of sociotechnical transformations to sustainability. *Current Opinion in Environmental Sustainability, 49*, 143–152. https://doi.org/10.1016/j.cosust.2021.04.010

Bijker, W. E. (1987). The social construction of Bakelite: Toward a theory of invention. In W. E. Bijker, T. Hughes, & T. Pinch (Eds.), *The social construction of technological systems* (pp. 159–187). MIT Press.

Castellacci, F., & Lie, C. M. (2017). A taxonomy of green innovators: Empirical evidence from South Korea. *Journal of Cleaner Production, 143*, 1036–1047. https://doi.org/10.1016/j.jclepro.2016.12.016

Dauvergne, P. (2023). Governing plastics: The power and importance of activism in the global South. *Environmental Science and Policy, 147*(April), 147–153. https://doi.org/10.1016/j.envsci.2023.06.011

Dauvergne, P., Iris, J., Beaudoin, S., Carney, B., Clapp, J., Cowan, E., De Groot, B., Farrelly, T., De Miranda, N., Mah, A., Mendenhall, E., Paik, R., Ralston, R., Stoett, P., Taggart, J., Tiller, R., Villarrubia-g, P., & Vince, J. (2025). Competing axes of power in the global plastics treaty: Understanding the politics of progress and setbacks in negotiating a high-ambition agreement. *Marine Policy, 181*, 106820.

Folino, A., Pangallo, D., & Salvatore, P. (2023). Assessing bioplastics biodegradability by standard and research methods: Current trends and open issues. *Journal of Environmental Chemical Engineering, 11*, 109424. https://doi.org/10.1016/j.jece.2023.109424
Fone, M. (2023). Curious questions: Why do we call a waterproof rain coat a mackintosh? Country Life. https://www.countrylife.co.uk/luxury/style/curious-questions-why-do-we-call-a-waterproof-rain-coat-a-macintosh-252460.
Frias, J. P. G. L., & Nash, R. (2019). Microplastics: Finding a consensus on the definition. *Marine Pollution Bulletin, 138*, 145–147. https://doi.org/10.1016/j.marpolbul.2018.11.022
Geels, F. W. (2002). Technological transitions as evolutionary reconfiguration processes: A multi-level perspective and a case-study. *Research Policy, 31*(8–9), 1257–1274. https://doi.org/10.1016/S0048-7333(02)00062-8
Geels, F. W. (2004). From sectoral systems of innovation to socio-technical systems insights about dynamics and change from sociology and institutional theory. *Research Policy, 33*, 897–920. https://doi.org/10.1016/j.respol.2004.01.015
Global Plastic Laws. (2025). UN plastics treaty. https://www.globalplasticlaws.org/un-global-plastics-treaty
Glover, D. (2022). Affordances and agricultural technology. *Journal of Rural Studies, 94*(May), 73–82. https://doi.org/10.1016/j.jrurstud.2022.05.007
Inverizzi, N., York, E., Dréano, C., Kaşdoğan, D., Kenner, A., Khandekar, A., Okuna, A., Otsuki, G. J., Raman, S., Schutz, T., Vasen, F., & Windle, A. (2023). Innovationism across transnational landscapes. *Engaging Science, Technology, and Society, 9*(2), 1–8.
Irwin, A. (2023). STS and innovation: Borderlands, regenerations and critical engagements. *Engaging Science, Technology, and Society, 9*(2), 41–56.
Ishmael-Perkins, N., Raman, S., Metcalf, J., Strudwicke, I., Gascoigne, T., & Leach, J. (2023). The contextualization Deficit: Reframing Trust in Science for Multilateral Policy. https://doi.org/10.24948/2023.10.
Jänicke, M., & Jörgens, H. (2023). Ecological modernization and beyond. In H. Jörgens, C. Knill, & Y. Steinebach (Eds.), *Routledge handbook of environmental policy* (1st ed., pp. 109–137). Routledge. https://doi.org/10.4324/9781003043843-7
Jewitt, S., & Raman, S. (2017). Energy poverty, institutional reform and challenges of sustainable development: The case of India. *Progress in Development Studies, 17*(2), 173–185. https://doi.org/10.1177/1464993416688837
Joly, P. B. (2019). Reimagining innovation. In *Innovation beyond technology: Science for society and interdisciplinary approaches* (pp. 25–45). Springer Singapore.
Knight, L. (2014). A brief history of plastics, natural and synthetic. BBC. https://www.bbc.com/news/magazine-27442625
Knoblauch, D., & Mederake, L. (2021). Government policies combatting plastic pollution. *Current Opinion in Toxicology, 28*, 87–96. https://doi.org/10.1016/j.cotox.2021.10.003
Köhler, J., Geels, F. W., Kern, F., Markard, J., Onsongo, E., Wieczorek, A., Alkemade, F., Avelino, F., Bergek, A., Boons, F., Fünfschilling, L., Hess, D., Holtz, G., Hyysalo, S., Jenkins, K., Kivimaa, P., Martiskainen, M., Mcmeekin, A., Susan, M., et al. (2019). An agenda for sustainability transitions research: State of the art and future directions. *Environmental Innovation and Societal Transitions, 31*, 1–32. https://doi.org/10.1016/j.eist.2019.01.004
Landrigan, P. J., Dunlop, S., Treskova, M., Raps, H., Symeonides, C., Muncke, J., Spring, M., Stegeman, J., Almroth, B. C., Chiles, T. C., Cropper, M., Deeney, M., Fuller, L., Geyer, R., Karasik, R., Mafira, T., Mangwiro, A., Matias, D. M., Mulders, Y., et al. (2025). The Lancet Countdown on health and plastics Philip. *The Lanclet, 406*, 1044–1062. https://doi.org/10.1016/S0140-6736(25)01447-3
Leach, M. (2015). What is green? Transformation imperatives and knowledge politics. In I. Scoones, M. Leach, & P. Newell (Eds.), *The politics of green transformations* (1st ed., pp. 25–38). Routledge. https://doi.org/10.4324/9781315747378-2
Levidow, L., & Raman, S. (2020). Sociotechnical imaginaries of low-carbon waste-energy futures: UK techno-market fixes displacing public accountability. *Social Studies of Science, 50*(4), 609–641. https://doi.org/10.1177/0306312720905084

Li, P., Wang, X., Su, M., Zou, X., Duan, L., & Zhang, H. (2021). Characteristics of plastic pollution in the environment: A review. *Bulletin of Environmental Contamination and Toxicology, 107*(4), 577–584. https://doi.org/10.1007/s00128-020-02820-1

Macleod, M., Arp, H. P. H., Tekman, M. B., & Jahnke, A. (2021). The global threat from plastic pollution. *Science, 65*(7491), 61–65.

Masco, J. (2017). The crisis in crisis. *Current Anthropology, 58*, S65–S76. https://doi.org/10.1086/688695

Miller, C. A. (2023). STS and the design of futures. In Z. Baker, T. Law, M. Vardy, & S. Zehr (Eds.), *Climate, science and society* (pp. 187–194). Routledge.

Mohr, A., & Raman, S. (2013). Lessons from first generation biofuels and implications for the sustainability appraisal of second generation biofuels. *Energy Policy, 63*, 114–122. https://doi.org/10.1016/j.enpol.2013.08.033

Mol, A. P. J., Spaargaren, G., & Sonnenfeld, D. A. (2020). Ecological modernisation: Three decades of policy, practice and theoretical reflection. In A. P. J. Mol, D. A. Sonnenfeld, & G. Spaargaren (Eds.), *The ecological modernisation reader: Environmental reform in theory and practice* (pp. 3–14). Routledge.

Nelms, S. E., Easman, E., Anderson, N., Berg, M., Coates, S., Crosby, A., Eisfeld-pierantonio, S., Eyles, L., Flux, T., Gilford, E., Giner, C., Hamlet, J., Hembrow, N., Hickie, J., Hopkinson, P., Jarvis, D., Kearsley, J., Millard, J., Nunn, F., et al. (2022). The role of citizen science in addressing plastic pollution: Challenges and opportunities. *Environmental Science and Policy, 128*, 14–23. https://doi.org/10.1016/j.envsci.2021.11.002

Pfotenhauer, S., Laurent, B., Papageorgiou, K., & Stilgoe, J. (2022). The politics of scaling. *Social Studies of Science, 52*(1), 3–34. https://doi.org/10.1177/03063127211048945

Reynolds, D. O. (2025). Avoiding plastics at the household level: Household agency as a mechanism in transitions. *Environmental Innovation and Societal Transitions, 56(March), 100983*. https://doi.org/10.1016/j.eist.2025.100983

Rodner, L. (Ed.). (2004). *The technological fix: How people use technology to create and solve social problems*. Routledge.

Rosenboom, J., Langer, R., & Traverso, G. (2022). Bioplastics for a circular economy. *Nature Reviews Materials, 7*, 117–137. https://doi.org/10.1038/s41578-021-00407-8

Rujnić-Sokele, M., & Pilipović, A. (2017). Challenges and opportunities of biodegradable plastics: A mini review. *Waste Management & Research: The Journal for a Sustainable Circular Economy, 35*(2), 132–140. https://doi.org/10.1177/0734242X16683272

Santos, D. (2023). Conceptualizing the democratization of innovation through transitions theory: A case study of biohacking in community science labs. *Environmental Innovation and Societal Transitions, 49*, 100783. https://doi.org/10.1016/j.eist.2023.100783

Sarewitz, D., & Nelson, R. (2008). Three rules for technological fixes. *Nature, 456*, 871–872.

Schmaltz, E., Melvin, E. C., Diana, Z., Gunady, E. F., Rittschof, D., Somarelli, J. A., Virdin, J., & Dunphy-daly, M. M. (2020). Plastic pollution solutions: Emerging technologies to prevent and collect marine plastic pollution. *Environment International, 144*(August), 106067. https://doi.org/10.1016/j.envint.2020.106067

Schofield, J., Leterme, S., Macgregor, M., Snigirova, A., & Werner, A. D. (2025). Archaeological and systemic context for the plastic age: Theorising the formation of contemporary and future archaeological records. *Cambridge Prisms: Plastics, 3*, e33. https://doi.org/10.1017/plc.2025.10028

Schot, J., & Steinmueller, W. E. (2018). Three frames for innovation policy: R&D, systems of innovation and transformative change. *Research Policy, 47*(9), 1554–1567. https://doi.org/10.1016/j.respol.2018.08.011

Sengers, F., Wieczorek, A. J., & Raven, R. (2019). Experimenting for sustainability transitions: A systematic literature review. *Technological Forecasting and Social Change, 145*, 153–164. https://doi.org/10.1016/j.techfore.2016.08.031

Shortall, O. K., Raman, S., & Millar, K. (2015). Are plants the new oil? Responsible innovation, biorefining and multipurpose agriculture. *Energy Policy, 86*, 360–368. https://doi.org/10.1016/j.enpol.2015.07.011

Shove, E., Trentmann, F., & Watson, M. (2018). Introduction – infrastructures in practice: The evolution of demand in networked societies. In E. Shove & F. Trentmann (Eds.), *Infrastructures in practice: The dynamics of demand in networked societies* (1st ed., pp. 1–5). Routledge.

Stallard, E., & Poynting, M. (2025). Global plastic talks collapse as countries remain deeply divided. BBC. https://www.bbc.com/news/articles/cvgpddpldleo

Steffen, W., Richardson, K., Rockström, J., Cornell, S. E., Fetzer, I., Bennett, E. M., Biggs, R., Carpenter, S. R., De Vries, W., De Wit, C. A., Folke, C., Gerten, D., Heinke, J., Mace, G. M., Persson, L. M., Ramanathan, V., Reyers, B., & Sörlin, S. (2015). Planetary boundaries: Guiding human development on a changing planet. *Science, 347*(6223), 1259855. https://doi.org/10.1126/science.1259855

Stirling, A. (2011). Pluralising progress: From integrative transitions to transformative diversity. *Environmental Innovation and Societal Transitions, 1*(1), 82–88. https://doi.org/10.1016/j.eist.2011.03.005

Takalo, S. K., Sayyadi, H., & Shahabaldini, Z. (2021). Green innovation: A systematic literature review. *Journal of Cleaner Production, 279*, 122474. https://doi.org/10.1016/j.jclepro.2020.122474

UNEP. (2025). About the UNEP Plastics Initiative. https://www.unep.org/topics/plastic-pollution/one-plastics-initiative/about-unep-plastics-initiative

Wyatt, S. (2007). Technological determinism is dead; long live technological determinism. In E. J. Hackett, O. Amsterdamska, M. E. Lynch, & J. Wajcman (Eds.), *Handbook of science and technology studies* (3rd ed., pp. 165–180). The MIT Press.

Zalasiewicz, J., Waters, C. N., Ivar, J. A., Corcoran, P. L., Barnosky, A. D., Cearreta, A., Edgeworth, M., Ga, A., Jeandel, C., Leinfelder, R., Mcneill, J. R., Steffen, W., Summerhayes, C., Wagreich, M., Williams, M., Wolfe, A. P., & Yonan, Y. (2016). The geological cycle of plastics and their use as a stratigraphic indicator of the Anthropocene. *Anthropocene, 13, 4–17.* https://doi.org/10.1016/j.ancene.2016.01.002

Sujatha Raman is a Professor and Director of the Australian National Centre for the Public Awareness of Science (CPAS), Australian National University (ANU), and UNESCO Chair-holder in Science Communication for the Public Good. She is an interdisciplinary social scientist interested in how matters of sustainability and equity intersect, and what these intersections mean for re-imagining how we bring science and innovation to bear on the task of building better planetary futures. She has published widely on responsible innovation and emerging technology assessment, the science/policy interface, and energy transitions and their material underpinnings. Before joining ANU in 2018, she was Co-director of the Institute for Science and Society (ISS), University of Nottingham (UK), and Director of the Leverhulme Research Programme, '*Making Science Public*' which spanned the Universities of Nottingham, Sheffield, and Warwick.

Dan Santos is a Research Fellow at the Australian National Centre for the Public Awareness of Science (CPAS) at the Australian National University, working with the UNESCO Chair in Science Communication for the Public Good. His current research interests revolve around the political economic and public engagement dimensions of emerging technologies, especially concerning questions around openness and innovation. On these topics, he has conducted empirical research on synthetic biology and stem cells, and his work engages with and contributes to the fields of science and technology studies (STS), science communication, and economic geography. He received his PhD in Geography from Clark University.

Chapter 2
Japanese Views of Nature: Relationships with Science, Technology, and Innovation, and Responsibility

Kei Kano

Abstract This chapter first explains the Japanese approach to adopting Western technology: *Wakon-Yosai* (a combination of Japanese traditions and Western skills). This approach involves adopting Western technology while not westernising the traditional and philosophical foundations of Japanese culture. Next, I introduce case studies based on *Wakon-Yosai* and examine its limitations. For instance, innovative start-up products such as Takadiastase and Ajinomoto have been established, grounded on traditional Japanese knowledge related to *koji* mold umami. Moreover, it is believed that Japan's first Nobel laureate, Hideki Yukawa, formulated a theory pertaining to quantum mechanics, drawing upon the Cheng–Zhu School and Western science. Conversely, it is noted that the Japanese views of nature has not contributed to the advancement of seismology. Furthermore, this chapter highlights Japanese views of nature in contrast to Western science and also examines the sense of responsibility in Japan, which differs from the West. It is crucial to pay attention to the differences in the sense of responsibility brought about by different views of nature when considering responsibility for environmental issues. Finally, three perspectives—integrating Western science with Japanese views of nature, promoting environmental responsibility and harmony between humans and nature, and avoiding unnecessary intervention while maintaining responsible research and innovation—are proposed to achieve a genuine integration of Japanese views of nature with Western science, technology, and innovation beyond *Wakon-Yosai*, by utilising Kakishibu as a case study.

Keywords Japanese views of nature · Science, Technology and Innovation · Responsibility · *Wakon-Yosai* · Western science · Kakishibu · Philosophy

K. Kano (✉)
Faculty of Education, Shiga University, Otsu, Shiga, Japan

The Australian National Centre for the Public Awareness of Science, Australian National University, Canberra, ACT, Australia
e-mail: kkano@edu.shiga-u.ac.jp; Kei.Kano@anu.edu.au

K. Kano et al. (eds.), *Traditional Knowledge for Green Innovation with Japanese Persimmon Tannin, Kakishibu*, https://doi.org/10.1007/978-981-95-7805-4_2

2.1 Japan's Approach to Adopting Western Technology: *Wakon-yosai* (a Combination of Japanese Traditions and Western Skills)

This section provides an overview of Japan's approach to adopting Western technology: *Wakon-Yosai* (a combination of Japanese traditions and Western skills) and the historical and cultural background that shaped it. *Wakon-Yosai* is unique to Japan, which has separated metaphysics from physical sciences and adopted only the physical sciences of the West. Catching up with and surpassing Western science and technology were among the goals. Additionally, the University of Tokyo, Japan's first university chartered in 1877, is recognised for pioneering the concept of a university organised by discipline, predating the world's leading institutions, based on the principles of *Wakon-Yosai*. Thus, *Wakon-Yosai* is a crucial concept that has influenced not only the development of science and technology in Japan but also higher education that supports it.

2.1.1 History of Interactions between Japan and Western Technology: from the Introduction of Firearms to the Meiji Restoration

The encounter between Japan and Western technology is often traced back to the introduction of firearms in 1543. A ship carrying Portuguese traders to Ningbo, a Chinese trading port, was washed ashore on Tanegashima, an island in present-day Kagoshima Prefecture. The Portuguese demonstrated their matchlock guns, marking Japan's first encounter with firearms. The lord of Tanegashima, Tokitaka Tanegashima, quickly bought two matchlock guns and told his chief blacksmith, Yaita Kinbei Kiyosada, to make copies of them. Initially, he had trouble making the screw hole that blocks the tail of the gun barrel. However, he learned these techniques from the Portuguese who visited the island the following year, and within 8 months of the guns' introduction, he had successfully produced domestic firearms. These domestically made guns, known as *Tanegashima*, spread among Japan's feudal warlords and became important weapons. As this story of the introduction of firearms shows, the speed at which Japan was able to copy and produce similar models of this Western technology was impressive. This also shows the high standard of blacksmithing skills in Japan at that time.

The next major event in Japan's interaction with Western technology was *sakoku* (closed country policy), which lasted from 1639 to 1854, spanning 215 years. This policy had a significantly harmful effect on the interaction between Japan and Western technology. The *sakoku* policy developed through a series of measures, including the banning of Christianity in 1613 and the restriction on overseas travel by Japanese citizens and their return from abroad in 1635, gradually limited foreign trade to the Netherlands and China from 1639. As a result, contact with the West was restricted to the Netherlands in the confined area of Dejima in Nagasaki.

The import of Western books was also banned, but in 1720, the eighth Shogun Yoshimune Tokugawa lifted the ban on Western books unrelated to Christianity. This led to the spread of Dutch learning schools, which aimed to accept Western scholarship and technology through the Dutch language. Interestingly, Yoshimune Tokugawa is also known for bringing the first elephant to Japan, which shows his interest in engaging with the outside world. Later, in 1774, Genpaku Sugita, Ryotaku Maeno, and others published *Kaitai Shinsho,* Japan's first full-scale translation of a Western scientific text. The original work was Anatomische Tabellen, a medical book by the German physician Johann Adam Kulmus, and its Dutch translation was used as the base text. This was the first comprehensive translated medical book in Japan, and it is thought that Japanese terms for nerves (*Shinkei*), cartilage (*Nankotsu*), and arteries (*Domyaku*) were developed during this period. The publication of *Kaitai Shinsho* triggered the translation of Dutch books in various fields such as astronomy and geography into Japanese. In this sense, the publication of *Kaitai Shinsho* can be regarded as a significant milestone in the adoption of Western science and technology in Japan, following the introduction of firearms.

However, it is not to say that there was no pressure against studying Dutch, which aimed to adopt Western technology under the *sakoku* policy. In 1828, the German doctor Philipp Franz von Siebold was expelled from Japan for trying to take out of Japan *Dai Nihon Enkai Yochi Zenzu*, maps of Japan's Coastal Area (known as Ino's Map), which was banned for national security reasons. This incident, known as the Von Siebold Incident, resulted in the punishment of several Dutch scholars who had been taught by Siebold, by the shogunate. In 1839, Dutch scholars Choei Takano and Kazan Watanabe were arrested for criticising the *sakoku* policy in an incident known as the *Bansha no Goku* (indictment of the Society for Western Study). This was seen as a crackdown on freedom of speech regarding Dutch study, and one reason for this was that Confucianism, especially the Cheng–Zhu School, which is one of the main strands of neo-Confucianism, was considered the orthodox scholarship by the shogunate at the time. More broadly, these episodes suggest a recurring tension under *sakoku*: Western learning was tolerated insofar as it produced practical knowledge, while open debate and critical inquiry were treated as politically risky. Watanabe (2000) characterises this as selective adoption: pursuing Western scholarship while constraining the spirit of free and critical thinking that enables it.

Nevertheless, Japan's *sakoku* policy was not perpetual. The event known as the Arrival of the Black Ships in 1853 resulted in the cessation of this policy. In 1852, American Matthew Calbraith Perry, who had been appointed as the admiral of the East India Squadron, arrived unexpectedly in Uraga with two steamships and two sailing ships, equipped with gun ports. The circumstances during the period when he advocated for the opening of the country to the Edo shogunate are preserved in the comic tanka playing on *jokisen* (steamships/high-grade green tea), conveying the shock of Perry's arrival and the sleeplessness it provoked. Subsequently, he made another visit in 1854, and after approximately a month of negotiations, the Japan–US Treaty of Peace and Amity was finalised, terminating the *sakoku* policy.

During the negotiations, demonstrations of telegraph machines and steam engine models were conducted. Perry's official expedition account describes how the Japanese reaction to these demonstrations was one of profound astonishment, portraying the encounter as a dramatic revelation to a people only partly acquainted with the achievements of science and industry. With the advent of firearms, blacksmiths were capable of producing them through imitation; however, advancements in technologies such as electricity and steam necessitated the mastery of all aspects, from basic science to application (Watanabe, 2000). As depicted on the cover of Western Affairs by Yukichi Fukuzawa (1866), which features illustrations of steamships, steam locomotives, and telegraph lines encircling the globe, it is evident that steam power and electricity have been pivotal forces propelling Western civilisation forward across diverse domains, including industry, transportation, communication, and military affairs (Watanabe, 2000).

Consequently, in 1868, the Meiji government formulated one of the oaths in the *Gokajo no Goseimon* (Oath in Five Articles) stating "Knowledge shall be sought throughout the world so as to strengthen the foundation of imperial rule." This initiative was intended to actively integrate Western science and technology.

2.1.2 The Adoption of Western Science and Technology During the Meiji Period: Study Abroad Programs and the Recruitment of Foreign Experts

During the Meiji era, there were two principal methods for adopting Western technology: pursuing studies abroad in Europe and America, and employing foreign experts. This section concentrates on the strategy of integrating Western science and technology while preserving *Wakon-Yosai*. Nonetheless, it is significant from the standpoint of the synthesis of Western and Japanese perspectives of nature—one of the primary themes of this chapter—that research focusing on phenomena and objects within Japan by hired foreigners, such as seismology conducted by British teachers J. Milne and A. Ewing, research on the sake brewing process by the British chemist R. W. Atkinson, and studies on *Makyo*, Japanese magic mirror, by W. E. Ayrton and J. Perry, was undertaken (Watanabe, 1991).

Numerous examples of studying abroad in Europe and America exist; however, as the intention is not to provide a comprehensive overview, I would like to focus on the case of Kenjiro Yamakawa studying in the United States. He was the first Professor of Physics at the University of Tokyo, the first university which adopted a distinctive faculty system at that time. Furthermore, he served as the university's president on two separate occasions.

Kenjiro Yamakawa was born in 1854, coinciding with the conclusion of Japan–US Treaty of Peace and Amity. He received an education predominantly emphasising samurai-era moral philosophy, with a particular focus on Cheng–Zhu School. He studied English and mathematics during the turbulent period of transition from the Edo period to the Meiji era. Following the advent of the Meiji era, he received directives from the government, and in 1871, he enrolled at the Sheffield Scientific School at Yale University, the predecessor institution of Yale University's Faculty of Science. During his time abroad, he pursued studies in physics and subsequently returned to Japan in 1875. Following his return, he was appointed as the first Japanese professor of physics at the University of Tokyo in 1879, overseeing the pioneering X-ray experiments within Japan. From 1887 onwards, he transitioned his focus towards educational administration, serving as the first president of Kyushu University in 1911, the president of Kyoto University in 1915, and holding the position of president at the University of Tokyo on two occasions, in 1901 and 1913. This is a concise overview of his career; however, I would like to highlight two episodes associated with *Wakon-Yosai*. Firstly, he was born during the Edo period when Christianity was prohibited, and he did not set foot in a church for 5 years because he believed that Christianity was heresy. He later reflected that in retrospect, it seems that it was not necessary to go that far, but it can be inferred that he was influenced by the social conditions of the time (Watanabe, 1991). The other is that he did not consider it necessary to educate men and women equally (Watanabe, 1991). The influence exerted by an individual who has served as the president of the University of Tokyo, Kyoto University, and Kyushu University can be considered substantial. Incidentally, as of 2019, the proportion of female students entering university in Japan in STEM fields is the lowest among OECD member countries (OECD, 2022).

2.1.3 Wakon-Yosai

Japan has incorporated Western science and technologies such as firearms, anatomy, telegraphs, steamships, steam locomotives, and X-rays. These instances exemplify the concept of *Wakon-Yosai* (a combination of Japanese traditions and Western skills). Simultaneously, it represented an importation of Western science and technology whilst preserving traditional Japanese ideas and cultural values. This included the prohibition of Christianity, the suppression of discourse by Cheng–Zhu School scholars against Dutch scholars, Kenjiro Yamakawa's abstention from visiting the church during his studies in the United States, and discriminatory notions regarding women's education. These are examples of the Japanese spirit part of *Wakon-Yosai*.

In other words, *Wakon-Yosai* can be understood as "Japan's approach to the West, attempting to differentiate between metaphysics and physical sciences, and exclusively embracing science and technology as the latter" (Murakami, 1977). In this reading, it operated as a strategy of modernisation that adopted Western science and technology while maintaining a distinct cultural and moral order. Conversely, Murakami (1977) also notes that *Wakon-Yosai* was founded on the misconception that "it is possible to separate Western metaphysics and physics." He cautions that such a separation may be over-optimistic: scientific practices are entangled with institutions, norms of inquiry, and broader intellectual commitments that do not necessarily remain outside the imported technology.

An illustrative example is the system design implemented at the inception of the University of Tokyo. The university was the pioneering institution worldwide to commence with departments that prioritised specialised fields from the beginning. Retrospectively, this model is often regarded as effective for rapidly narrowing the gap with the Western science and technology in specialised fields; however, it may also have encouraged forms of expertise that are strong in science and technology but weaker in cultural and historical literacy across traditions, as well as in broader international understanding (Watanabe, 2000). Simultaneously, Watanabe (1991) further argues that this "limited outlook" could obstruct a fuller grasp of Western civilisation and, consequently, of modern science as an outcome of human intellectual and spiritual activity.

What, then, constituted the Japanese spirit, namely, traditional Japanese thought and culture? The scope of traditional Japanese thought and culture is extensive, and it cannot be comprehensively addressed within the confines of this chapter. Therefore, I would like to focus on Japanese views of nature, which is intricately connected to Kakishibu and traditional knowledge, the core themes of this book. Furthermore, the scope of Japanese views of nature is extensive. As this publication pertains to science, technology, and innovation, I shall examine it in comparison with Western science. Further details will be provided in Sect. 2.3. However, here, I will present an overview of Japanese views of nature by offering some examples. Watanabe is a leading figure in this field, and numerous examples have been included in many of his publications. This overview is primarily derived from those works by Watanabe.

The first example is a proverb contrasts Eastern and Western medicine: "Eastern medicine heals the patient, Western medicine heals the disease." In Western practices, diseases and pathological sites have been objectified and addressed, whereas in Japan, humans have been perceived holistically, and medicine has been regarded as a philanthropic pursuit intended to save lives.

The second example pertains to traditional *ikebana*, a Japanese flower arrangement. According to Watanabe (1991), traditional *ikebana* arranges three branches to symbolise heaven, human being, and earth, while presenting them as if they originate from a single branch. This compositional logic foregrounds harmony and relationality. Watanabe (1991) further suggests that, unlike the European model—where tensions and oppositions are taken for granted—humans are understood as inseparable from, and integrated within, a harmonious universe.

The third example is presented in the lecture titled Zen Buddhism and Psychoanalysis by Suzuki (1960), where he compares a waka composed by Matsuo Basho with a poem authored by Alfred Tennyson. Alfred Tennyson's Flower in the Crannied Wall frames the flower as something to be extracted ("I pluck you out… root and all") and held as an object whose essence might be understood through analysis.

Building on Watanabe's reading (1995), Basho's opening phrase *yoku mireba* (looking carefully) can be interpreted as a mode of observation in which the boundary between subject and object is less sharply drawn, emphasising coexistence rather than extraction. This contrast is not merely literary but epistemic: Tennyson's stance aligns with an objectifying mode of inquiry that separates nature from the observer, whereas Basho's attention is non-interventionist and relational.

The fourth example is Sesshu (1474)'s landscape painting, *Shihombokugasansuizu*. The artificial elements of a small boat, a bridge, and the roof of a house, along with the person crossing the bridge, blend into nature. Apart from the slightly linear lines of the sail and roof, there are no other geometric shapes present. It effectively represents traditional understanding of the relationship between nature and humans within it (Watanabe, 1995).

The final example is the opening of *Hojoki* by Kamo no Chomei, which every Japanese has recited at least once. The passage frames nature—and human life within it—as ceaselessly changing, using the imagery of a river whose flow never stops and foam that gathers only to vanish. A brief excerpt captures the core intuition: "the flood ever changeth … the foam … vanishing, stayeth not."

According to Watanabe (1995), this opening succinctly articulates Japanese views of nature in which change is not exceptional but fundamental. Each example exhibits elements of Japanese views of nature: a holistic approach, a harmonious relationship between humans and nature, the absence of a clear distinction between subject and object, a blurred distinction between artificial and natural, and the impermanence of nature.

2.2 Case Studies: Innovations and Discoveries Based on Japanese Traditional Knowledge and Views of Nature

In this section, five examples of innovations and discoveries attributed to Japanese individuals, founded upon Japanese traditional knowledge and views of nature will be presented. Instead of engaging extensively with each example, I kindly ask for readers' understanding as concentrating on the aspects where Japanese traditional knowledge and views of nature significantly manifested. The initial two examples are cases where Western science and technology was effectively integrated with traditional knowledge, thereby fostering innovation. The subsequent two examples are cases of novel discoveries rooted in Japanese views of nature. The two individuals depicted in the initial two examples, as outlined in Sect. 2.1, pursued knowledge on an international scale and engaged in international study programs, whereas the two individuals presented in the subsequent examples did not partake in studying abroad. It is challenging to ascertain the precise impact of studying abroad on these cases; however, I would like to observe that there exists a noticeable contrast regarding the presence or absence of international study experience among these four instances. Furthermore, given that the primary focus of this book pertains to Kakishibu, I would like to include an example of tannin in green tea, which is traditionally recognised in Japan. Furthermore, as demonstrated in Sect.

2.1, there existed a discriminatory attitude towards women's education rooted in the ideology of *Wakon-Yosai*. This serves as a reminder that exemplary cases from that period were biased in favour of men. I appreciate readers' understanding on this matter in advance.

Finally, an example from the field of seismology, which was primarily developed by Westerners who arrived in Japan, will be discussed. This example demonstrates that earthquakes had not been studied from Japanese views of nature.

2.2.1 Drug Discovery from Sake Koji *by Jokichi Takamine*

Traditional Japanese sake brewing was designated as an intangible cultural heritage by UNESCO in December 2024. This case presents a pioneering start-up in Japan, founded by Jokichi Takamine (1854–1922), which specialises in the application of the *koji* mold utilised in traditional Japanese sake brewing. I will present his story of discovery and innovation from the Jokichi Takamine Collection compiled by Suzuki (2022) and the official webpage of the Jokichi Takamine Research Society.

Similar to Yamakawa, who was introduced in Sect. 2.1, Takamine was among those who undertook studies abroad during the Meiji period. He attended the University of Glasgow in the United Kingdom for 3 years starting in 1880. Upon returning to Japan, he argued that Japan's distinctive chemical industries were poorly understood in the West and therefore needed to be developed and led by Japanese actors (Suzuki, 2022). Amongst numerous candidates for Japanese-led manufacturing such as sake, soy sauce, miso, oil, salt, sugar, wax, paper, ink, pottery, lacquer, paint, dye, and medicine, he prioritised the manufacturing of sake. He also maintained that chemistry-related scholars should pursue such research not only for commercial returns but also for government revenue (Suzuki, 2022). He criticised the prevailing tendency to imitate Western practices under *Wakon-Yosai* and urged Japanese researchers to publish original work and demonstrate their own contributions (Suzuki, 2022).

To embody this, he conducted research and development to utilise *koji*, a substance used in sake brewing, as a substitute for malt traditionally employed in Western brewing. Consequently, he secured patents in the United Kingdom in 1887, France and Belgium in 1888, and in the United States in 1889. He subsequently travelled to the United States with the objective of expanding his business within a whiskey enterprise. It was anticipated that he could produce whiskey more cost-effectively and efficiently than traditional malt methods; however, he encountered substantial resistance from the malt industry, which advocates for conventional whiskey production, leading him to relinquish his efforts and return to Japan.

Following his return to Japan, he redirected his efforts from utilising the starch-converting capacity of *koji* in sake fermentation to supporting human digestion, leading to the development of Takadiastase, a digestive medicine for starch digestion. When it was marketed by Parke-Davis (subsequently incorporated into Pfizer) in 1895, it gained widespread popularity not only within the United States but also across Europe, serving individuals ranging from infants to adults. In Japan, he entrusted the sales rights to Matasaku Shiobara, which subsequently led to Daiichi Sankyo Co., Ltd., one of Japan's premier pharmaceutical companies. Furthermore, Takamine emphasised the importance of studying Eastern materials and spearheaded the development of the RIKEN Institute, which is currently known as Japan's largest comprehensive research institution renowned for high-quality research in a diverse range of scientific disciplines, from this conceptual foundation.

The success of Takadiastase is widely recognised as Japan's pioneering start-up case. This constitutes an exemplary illustration of how naturally occurring ingredients present in *koji*, a substance historically employed in traditional Japanese sake brewing, have contributed to the development of pharmaceuticals. Conversely, it is essential to acknowledge that it encountered opposition from conventional whiskey production practices in the United States. This highlights the challenge of respecting traditional techniques from other nations while also honouring one's own.

2.2.2 The Discovery of the Umami Substance and Entrepreneurial Endeavours by Kikunae Ikeda

Japanese cuisine was registered as an intangible cultural heritage by UNESCO in 2013. This case illustrates how Kikunae Ikeda (1864–1936) concentrated on the umami of kelp, an essential component in traditional Japanese soup stock preparation, which ultimately culminated in the founding of the company Ajinomoto. Soup stock is characterised as a broth abundant in umami, a fundamental element in Japanese culinary practices. Similar to Yamakawa and Takamine, Ikeda was among those who pursued studies abroad during the Meiji era. He conducted research at Ostwald's laboratory at the University of Leipzig in Germany for a period of one and a half years commencing in 1899. Upon returning to Japan, he suggested to the Commission on Isotopic Abundances and Atomic Weights that the atomic weight standard be established as Oxygen-16, a proposal that was subsequently adopted. However, he later shifted his focus to umami research (Kono, 2020).

The most pivotal turning point transpired in 1907, when Ikeda's spouse acquired a quantity of kelp intended for use as soup stock. Subsequently, he developed an interest in the potential for artificially synthesising the umami flavour characteristic of kelp. At that period, only four tastes were acknowledged: sour, sweet, salty, and bitter, and umami was an emerging concept. His enduring aspiration is to develop a flavourful and cost-effective seasoning with the aim of improving the nation's suboptimal nutritional status by transforming nutritious yet unpalatable foods into enjoyable cuisine. Additionally, it appears to have served as one of the inspirations for umami research (Fuke, 2008).

Following several experiments, in 1908, he isolated glutamic acid as an umami substance, which was initially designated as the taste of glutamic acid rather than umami. In the same year, he was awarded a patent for the production of glutamate salt used as a seasoning derived from hydrolysed wheat. Following numerous developments, the product was ultimately launched in 1909 under the name *Ajinomoto*. This resulted in the establishment of the current Ajinomoto Co., Ltd. He also contributed to the founding of the RIKEN Institute, which is currently known as Japan's largest comprehensive research institution renowned for high-quality research in a diverse range of scientific disciplines.

It was not until the discovery of the umami receptor in 2002 (Nelson et al., 2002) that umami was globally acknowledged as the fifth taste. The span of 94 years since the identification of the umami substance underscores a significant temporal distance. It serves as a reminder that, despite widespread recognition of certain knowledge at a local or traditional level, such understanding is frequently not acknowledged on the international stage unless it is substantiated with evidence aligned with Western scientific standards.

2.2.3 Establishment of Primatology by Kinji Imanishi

This case presents the objectives of Japanese primatology, as initiated by Kinji Imanishi (1902–1992), along with the traditional fieldwork methods that have been adopted. It is said that there is a new idea and academic framework aimed at deepening understanding of the origins of human society by clarifying the social structure of primates and their evolutionary history (Yamagiwa, 2009). The methodology encompasses habituation through feeding, individual identification, and long-term continuous observation (Matsuzawa, 2009). This has now become the global standard, but it was once regarded as a significant oversight in Western scientific methodology, which rigorously objectifies even humans (Murakami, 1977). Furthermore, he regarded nature as intrinsically harmonious, viewing species as fitting together within a larger organic whole, each finding its own niche (de Waal, 2003). Frans B. M. de Waal (2003) highlighted that the approach was innovative because it avoided a strict human–animal dualism, and from a Buddhist perspective, this is entirely plausible—perhaps even probable—and it carries no sense of insult.

This case illustrates how Japanese views of nature—seeing things holistically, harmoniously viewing humans and nature, not distinguishing between subject and object, blurring the distinction between artificial and natural, and seeing nature as impermanent—functioned effectively as a novel scientific method. As de Waal (2003) noted that Imanishi's approach represents a paradigm shift adopted across primatology and beyond. This case could illustrate a combination of Western traditions and Japanese skills, or Western version of *Wakon-Yosai*.

2.2.4 Hideki Yukawa's Nobel Prize–Winning Meson Theory

This case presents the meson theory advanced by Hideki Yukawa (1907–1981), the first Japanese Nobel laureate. It highlights the idea that Japanese views of nature may have influenced the development of this theory.

Hideki Yukawa addressed the question of why protons within the atomic nucleus, which ought to repel each other owing to electromagnetic forces, nevertheless stay in a relatively confined space. It was assumed that a significant interaction was taking place between neutrons and protons, which are also constituents of the nucleus. At that time, the latest theories proposed that electromagnetic and gravitational forces result from the exchange of massless particles, such as photons and gravitons. Yukawa proposed the concept that an exchange of a certain type of particle occurs between neutrons and protons, referring to this particle as a meson. The term meson was assigned because its calculated mass lies between that of an electron and a proton. Ultimately, between 1934 and 1935, he proposed theories concerning the strong force resulting from the exchange of mesons between protons and neutrons, predicted the existence of mesons, and so forth. Subsequently, in 1947, the π meson was discovered, and in 1949, he became the first Japanese to receive the Nobel Prize in Physics independently. The fundamental concept of meson theory can be expressed as p (proton) → n (neutron) + π (meson). There exists a possibility that Japanese views of nature influenced the process leading to the conception of this formula. As demonstrated by the formula, when a proton emits a π meson, it transforms into a neutron. Conversely, the equation n (neutron) + π (meson) → p (proton) indicates that a neutron which has received a meson is converted into a proton. Despite these changes, the overall structure involves the exchange of π mesons between protons and neutrons. Recalling Japanese views of nature discussed in Sect. 2.1, this overall and change aspect aligns with the characteristics of Japanese views of nature, which perceives phenomena holistically and regards nature as impermanent.

Conversely, it could be argued that this does not align with the Western conception of nature (Ito, 1995). As Ito (1995) argues that European thought has long sought immutable underlying substances, which can make particle transmutation harder to conceptualise. Yukawa contrasts Epicurean thought with Laozi and Zhuangzi, arguing that the latter begin from a standpoint where the distinction between being and non-being is not primary (Cai, 1992).

Laozi and Zhuangzi's philosophies originated in China. However, as indicated in Sect. 2.1.1, *Bansha no Goku*, the predominant school during the Edo period was the Cheng–Zhu School, which follows the principles of Laozi and Zhuangzi. In this context, it can be asserted that the philosophies of Laozi and Zhuangzi constituted one of the foundational elements of Japanese views of nature. As Ito (1995) cautions that this particular perspective was not necessarily deliberately employed in the development of Yukawa's meson theory. It remains merely a possibility; Ito (1995) suggests that continual proton–neutron transformation may have been easier to accept for Yukawa than for Europeans accustomed to immutable entities. This example illustrates what Yukawa characterised as the junction of the East and the West—a meeting of Eastern philosophy and Western physics grounded in modern rationalism (Cai, 1992). He also left the maxim that what is original is initially a minority, whereas majority tend not to be original," which would also be applicable to the previous example of Kinji Imanishi. The relationship between Eastern philosophy and Western science and technology demonstrates the potential for incorporation into original concepts, both historically and contemporaneously.

2.2.5 Discovery of Tannin in Green Tea by Michiyo Tsujimura

This case examines the discovery of vitamin C, catechins, and tannin in green tea by Michiyo Tsujimura (1988–1969), who was the first female agricultural doctor in Japan. This section's content heavily relies on the work conducted by the Institute for Gender Studies at Ochanomizu University (2003), which has compiled and disseminated accomplishments pertaining to Michiyo Tsujimura.

Tsujimura graduated from Tokyo Women's Higher Normal School (currently known as Ochanomizu University) in 1913. She subsequently attempted to enrol at Hokkaido Imperial University; however, she was not allowed to enroll as a regular student due to her gender (Institute for Gender Studies at Ochanomizu University, 2003). Consequently, she assumed the role of an unpaid assistant and engaged in research on silkworms. Following numerous developments, she became a member of the RIKEN, which is currently known as Japan's largest comprehensive research institution renowned for high-quality research in a diverse range of scientific disciplines, in 1924, and initiated her research on vitamin C in green tea. At that time, researcher H. S. Mitchell reported that Japanese green tea did not contain vitamin C, based on an extraction using boiling water for five minutes, although this conclusion was based on experiments conducted with extraction using boiling water for 5 minutes (Ochanomizu University Gender Research Center, 2003). In traditional Japanese green tea preparation, boiling water is typically not employed, which presents an impractical condition. When Tsujimura performed an experiment by extracting for 1 minute with water maintained at the appropriate temperature of 55–60 °c, it was demonstrated that green tea contains a substantial amount of vitamin C (Institute for Gender Studies at Ochanomizu University, 2003). Consequently, it is reported that the export volume of Japanese tea to North America experienced a significant increase (Institute for Gender Studies at Ochanomizu University, 2003).

Subsequently, she identified the astringent constituents in green tea, namely, catechins and tannins. The paper summarising these studies, On the Chemical Components of Green Tea, received high praise, and a doctoral degree in agriculture was awarded from Tokyo Imperial University (now the University of Tokyo), marking the inception of Japan's first female doctoral graduate in agricultural science (Institute for Gender Studies at Ochanomizu University, 2003).

This case illustrates the challenges and difficulties faced by female scientists within the framework of *Wakon-Yosai*. Additionally, it demonstrates that utilising local traditional knowledge, such as the traditional methods of brewing green tea, can contribute to the development of accurate scientific understanding.

2.2.6 Seismology Hardly Born from Japanese Views of Nature

Thus far, only positive examples based on Japanese traditional knowledge and Japanese views of nature have been introduced. However, the constraints imposed by the separation between metaphysics and physical sciences have been identified in Sect. 2.1.3. Therefore, I will examine the example of the establishment of seismology, a discipline now taken for granted in Japan, and consider why it was difficult to originate from Japanese views of nature.

As Watanabe (1991) argues, the absence of seismological tradition in Japan is not simply a matter of technical delay. Rather, it reflects a historically durable orientation toward natural calamities as existential and moral events—recorded, endured, and integrated into a broader sense of impermanence—rather than as phenomena to be isolated and investigated as objects of systematic inquiry. The following passage crystallises this contrast:

> Since ancient times, there have been innumerable earthquakes in Japan, yet it was not the Japanese who invented seismology. They behaved more like Kamo no Chomei, who in his *Sketches from a Small Hermit Cell* (Hojoki, 1212) enumerates natural catastrophes such as fires, earthquakes, typhoons, floods, drought, famines, and epidemics, reports of human fame and ruin, and describes the transience of the world, to find rest in the end in a lonely hut in the bosom of nature. "Nothing is better," wrote Kamo, "than to make friends solely with music and nature."

Two moves are central here: Watanabe links repeated earthquakes to a cultural response that privileges narration and resignation over investigation, and he uses Hojoki as an emblem of a worldview in which disaster becomes evidence of transience rather than a research problem.What matters for this chapter is the epistemic posture implied by Watanabe's contrast. In the Hojoki frame, earthquakes are intelligible as part of a world defined by flux and impermanence; the appropriate response is ethical and aesthetic attunement. Modern seismology, by contrast, presupposes that earthquakes can be abstracted from lived experience, rendered comparable across cases, and made legible through instruments, measurement, and institutionalised discussion. This difference echoes the distinction in Sect. 2.1 between relational, non-extractive observation (Basho) and an analytic stance that objectifies nature (Tennyson). Seen in these terms, the later importation of seismology during the Meiji period also illustrates a limit of *Wakon-Yosai*: separating metaphysics from physical sciences may accelerate the uptake of techniques, but it does not automatically supply the epistemic institutions and norms of inquiry through which a natural phenomenon becomes a stable object of research. Ultimately, advancements in seismology were facilitated by foreign experts who arrived during the Meiji era. Notably, three experts from the UK—John Milne, James Ewing, and Thomas Gray—who firsthand experienced a significant earthquake in 1880, contributed notably to this progress. They pioneered the development of the Ewing-Gray-Milne seismograph, a comprehensive seismic instrument, thus establishing the groundwork for contemporary seismology in Japan. Furthermore, in 1880, the world's first seismological society, known as the Seismological Society of Japan, was founded by Milne. It is reported that approximately 80 members (68%) of the original membership of 117 members were foreigners.

This case suggests that, despite their frequency in Japan, earthquakes were not traditionally framed as objects of systematic inquiry, a tendency that can be linked to Japanese views of nature discussed in Sect. 2.1.

2.3 Japanese Views of Nature and Sense of Responsibility

I have previously examined an overview of Japanese views of nature in Sect. 2.1, as well as case studies of innovations and discoveries that are either based on or not based on Japanese views of nature, as briefly delineated in Sect. 2.2. This section will elucidate Japanese views of nature in more detail and examine the associated sense of responsibility. However, as noted in Sect. 2.1.3, the scope of Japanese views of nature is broad, and given that this publication focuses on science, technology, and innovation, I will consider it in comparison with Western science. This account is not meant to imply that Japanese views of nature is static. Rather it is helpful to consider some distinctive features of the traditional view, even though this may co-exist with changing ideas of nature.

2.3.1 Japanese Views of Nature

In this section, I intend to elucidate Japanese views of nature by contrasting it with Western science. Here, I use "view of nature" to refer to a culturally situated process whereby sensory perception is integrated into unified images, accompanied by affective experience (pleasure, discomfort, joy, anger, sadness), retained in memory, and subjected to personal valuation—meaning and worth—through which desires to engage with nature arise and a will to realise those desires is cultivated (Suzuki, 2018).

This section will include numerous comparisons with Western science; however, I would like to highlight four precautions in accordance with the observations of Aikenhead and Ogawa (2007). The initial consideration pertains to the necessity of recognising the context-dependent nature of the subject matter to prevent the perpetuation of stereotypes resulting from excessive generalisation. The subsequent consideration involves the acknowledgement that the same English term may denote different concepts. For instance, Nature and *Shizen* are distinct concepts; however, both are translated as "Nature" in English. The translation issues of Nature and *Shizen* will be examined in detail later. The third consideration involves the necessity of selecting appropriate terminology tailored to the target audience. As will be elaborated subsequently, expressions such as Japanese views of nature, Japanese way of knowing nature, and Indigenous science will be employed interchangeably, each carrying distinct nuances. Our terminology facilitates the articulation of specific ideas in ways that are sensitive to different worldviews, thereby potentially enhancing reader comprehension. The fourth and final point concerns the inherent bias introduced by Western perspectives in our comparisons, a consequence of our use of English as the language of communication. Although we maintain that Western science and a Japanese views of nature are rooted in cultural contexts and that no single framework is inherently superior, our comparative framework is inevitably anchored within a paradigm that is implicitly embedded in the English language.

2.3.1.1 Relationship Between Humans and Nature

First and foremost, I would like to address the relationship between humans and nature. As Watanabe (1991) notes below, some Western traditions—particularly those shaped by Christian theological assumptions—have often framed humans and nature in a dichotomous relationship, distinguishing the human as subject from nature as object. This formulation is helpful here because it clarifies one influential pathway by which treating nature as an object of research and practical use became intellectually and culturally intelligible.

> God is the real source and starting point of the traditional Western conception of the world and nature. The world exists because God exists and everything in the universe is created by God. Over all this, humanity towers as a special creation because human beings are created in the image of God and they alone were the recipients of God's breath. This clearly distinguishes human beings from the rest of creation. Just as there is an absolute difference between creator and creation, so there exists also an absolute difference between the creators touched by God's breath and the rest of creation. As God rules over the whole creation, so human beings rule over all other creatures. From this viewpoint there results finally the attitude of people in the West toward nature, which they look at objectively and make into an object for research and practical use.

Although this chapter does not focus on the historical development of Western science, it is helpful to note several influential strands in how the human–nature relationship has been framed. Francis Bacon is often associated with a program of expanding human power over nature through empirical inquiry and technological arts—an orientation sometimes summarized as a project of "mastery" or "control" of nature. René Descartes, likewise, helped consolidate a mechanistic conception of the natural world and a dualistic framework that encouraged treating nature as an object of analysis and intervention. Such ambitions are clearly expressed in early modern scientific writing. In the preface to *Micrographia*, Robert Hooke argues that humans surpass other creatures not only by using nature to survive, but also by reflecting on it and reshaping it—comparing, modifying, and improving it for different purposes. With skill and experience, some people can sharpen this ability so much that they outdo others in observation and reasoning almost as much as humans outdo animals.

Immanuel Kant is often read as shifting the discussion from nature "in itself" to the conditions under which nature can be experienced and known: nature as phenomena is structured by the forms and concepts through which human cognition organises experience. Taken together, these perspectives have frequently been interpreted as supporting a tendency—prominent in some accounts of Western modernity—to distinguish human subjects from a natural world treated as an object of inquiry and use. Conversely, in discussions of a Japanese views of nature, humans and nature are often described as harmonious and closely intertwined, with less emphasis on a strict subject–object dichotomy (Fig. 2.1). Furthermore, nature has often been regarded as impermanent. These attributes are exemplified in the case studies presented in Sects. 2.1.3 and 2.2. What is the reason behind the relative lack of a sharp distinction between nature and humans within Japanese views of nature? I would like to explore this by considering how the modern Western concept of Nature came to be translated and conceptualised in Japan.

The Japanese term, now translated as nature, is known as *shizen*. However, *shizen* historically did not necessarily denote nature in the modern European sense. It functions as an adjective or adverb signifying *onozukara*, meaning "as it is" (Yanabu, 1977), or as a noun meaning *onozukara shikari*, meaning spontaneity (Suzuki, 2018). The term has been employed to denote the state of existing as it is and a state that has not been altered by humans. In Japan, it is frequently observed that being *shizen* is regarded as a good thing, indicating an ideal or desirable condition (Watanabe, 1995). For example, *shizen* food refers to organic food, signifying food that is beneficial due to its absence of pesticide use. *Shizen* energy means renewable energy. Furthermore, deriving from the interpretation of *shizen*, the term *shizen* traffic jam is employed to describe traffic congestion that occurs without an identifiable cause such as an accident or construction. As evidenced by the use of *shizen* despite the fact that organic food, renewable energy, and traffic congestion are all artificial, the distinction between nature and artificiality can remain ambiguous within Japanese views of nature (Watanabe, 1995).

One of the sources of the meaning of *shizen* can be identified within Laozi-Zhuangzi philosophy. Laozi regarded it as a state that exists naturally without human intervention, while Zhuangzi presented the concept of all things being equal, which signifies endeavouring to harmonise rather than oppose heaven and earth (Ito, 1999). The alchemy discussed in Taoism, which follows the lineage of

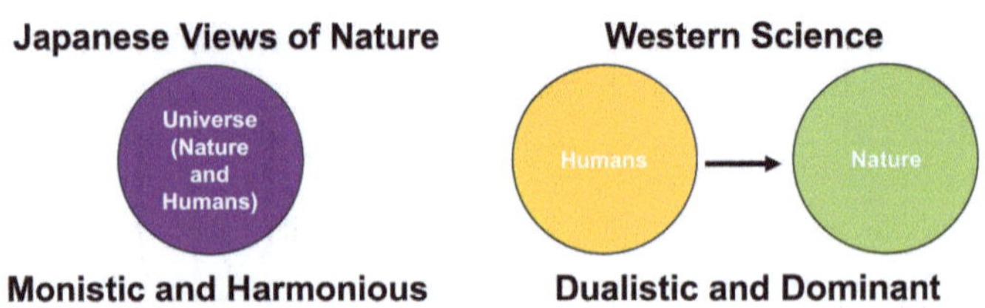

Fig. 2.1 Different views of Nature between Japanese views of nature and Western science

Laozi and Zhuangzi, was based on the idea that since changing is the nature of heaven and earth, it is possible to make gold from other things (Ito, 1999). This constitutes the perspective of impermanence previously referenced in Sect. 2.1.3, and as elaborated in Sect. 2.2.4. In the Cheng–Zhu School, which follows the lineage of Taoism, it is articulated that *li,* principle of organisation (Needham & Ling, 1956), is undifferentiated because it is shared by both nature and humans (Ito, 1999). Furthermore, *shizen* was adopted as a Buddhist term and was also referred to as *jinen*. Both *shizen* and *jinen* are represented by the same kanji, 自然. Given that the meaning of *jinen* is essentially identical to *shizen*, I shall not explore it further in this context.

As observed, *shizen* did not originally denote natural objects in an object-centred sense. The contemporary understanding of the term *shizen* was established as the translation of nature within the context of *Wakon-Yosai* during the Meiji era. During this period, *shizen* also came to signify natural objects. The attribution of the meaning of nature to *shizen* is often discussed as having taken shape in Japan and later circulating more widely in East Asia. In contemporary Japan and China, the term represented by the kanji or Chinese character for *shizen,* 自然, encompasses both the meaning of "as it is" and the concept of nature of Western origin.

2.3.1.2 Japanese Way of Knowing Nature

Having considered Japanese views of nature, I would now like to examine a way of knowing nature in Japanese culture by contrasting it with Western scientific paradigms. As observed in Sect. 2.3.1.1, the term employed in English is nature, yet it is imperative to acknowledge that it aligns with the Japanese concept of *shizen*. A Japanese way of knowing nature is often regarded as interchangeable with terms such as neo-Indigenous Knowledge, Indigenous science, traditional knowledge, traditional wisdom, traditional ecological knowledge, Native science, and a Japanese way of *seigyo-shizen* (Aikenhead & Ogawa, 2007); however, I will refrain from exploring the distinctions among these terms herein.

Based on the discussion by Aikenhead and Ogawa (2007), I will highlight a Japanese way of knowing nature (*shizen*) in contrast to Western science, as shown in the table below (Table 2.1). Here, "Western science" is used as a reader-friendly label for what Aikenhead and Ogawa (2007) call "Eurocentric sciences" (plural). Because Aikenhead and Ogawa (2007) emphasise substantial overlap between the Japanese way of knowing nature and Indigenous ways of living in nature (IWLN), Table 2.1 also incorporates selected IWLN characteristics where they plausibly represent shared features rather than uniquely Japanese traits. As previously noted by Aikenhead and Ogawa (2007), it is important to understand that this is not about stereotyping the two in comparison to Western science, but rather recognising that they are context-dependent. The table is simply a summary provided by the author for ease of reference. It is not the author's intention for the table solely to be copied and circulated out of context, which could promote stereotyping. Please be aware of this prior to proceeding.

Table 2.1 Japanese way of knowing nature and Eurocentric sciences

	Japanese way of knowing nature (*shizen*)*	Western Science (Eurocentric sciences in Aikenhead & Ogawa, 2007)*
Common grounds	Empirically based (sharing intellectual processes such as observing, looking for patterns, inferring, predicting, verifying, etc.) Elders and scientists exercise rational thoughts, and when doing so, they demonstrate communal characteristics (honesty, inquisitiveness, perseverance, and open-mindedness). Dynamic and historic, that is, change is based on past understandings Outcomes are continually being revised in light of new observations and new ideas Culture-based Typically less explicitly place-based; by contrast, Indigenous ways of living in nature is often described as strongly place-based	
Attribute of nature	Mysterious	Knowable
Fundamental value	Harmony with nature as a dynamic and multidimensional balancing of interrelationships for survival	Tends to emphasize power and domination over nature
Notion of time	Circular	Rectilinear
Outcome	Wisdom-in-action (models include symbols, numbers, geometric shapes, special objects, art forms, songs, stories, proverbs, metaphors, structures, and the always present circle)	Knowledge
Concept of ways of living/knowledge	Monist	Dualist
	Holistic (no separation of science, art, religion, philosophy, or aesthetics)	Reductionistic
	Verb-based	Noun-based
	Relational	Anthropocentric
Social goals	Wisdom-in-action for survival	Individual scientific credibility, corporate profits, medical advances, national security, economic progress, knowledge for its own sake, among many others
Intellectual goals	Co-existence with the mystery of nature by celebrating mystery through the maintenance of a host of relationships	Eradication of mystery by describing and explaining nature in ways familiar to a community of Eurocentric scientists
Culture-based observations	With the assumption (faith) that the observer is related and responsible in some way to what is being observed, and those observations are mainly qualitative	With the assumption (faith) that the observer and the observed are not at all related (i.e. the objective observer), what is being observed is more or less mathematical, quantitative observations are most often expected

(continued)

Table 2.1 (continued)

	Japanese way of knowing nature (*shizen*)*	Western Science (Eurocentric sciences in Aikenhead & Ogawa, 2007)*
Culture-based predictions	Harmonising with nature for the purpose of survival by maintaining a balance within a web of relationships, not harmonising with nature in any romanticised sense	Exercising power and dominion over nature
Culture-based empirical methodologies	Holistic and spiritual power	Power of reductionism and Cartesian dualism
Association with human action	Intimately, subjectively, morally, and ethically related to human action with respect to seven generations to come	Formally and objectively decontextualised from normative prescriptions of human action
Validity	Content validity as defined by Aristotle's notion of intelligible essences and supported by tens of thousands of years of survival based on that content Content validity concerns what the universe is	Predictive validity, the cornerstone of the epistemology of natural philosophy and Eurocentric sciences since the sixteenth century

Summarised from Aikenhead and Ogawa (2007)
*This table is a heuristic summary. The left column centres the Japanese way of knowing nature (incl. *seigyo-shizen*) in Aikenhead & Ogawa (2007), but it also deliberately includes IWLN (Indigenous ways of living in nature) features in areas where the paper suggests substantial overlap/common ground. Accordingly, entries in this column are not claims about all Japanese people/culture, nor should they be copied or circulated out of context in ways that risk stereotyping or essentializing. The label "Western science" is used here only as a conventional shorthand. Following Aikenhead & Ogawa (2007), the intended referent is "Eurocentric sciences" (plural), introduced to emphasize culturally situated and internally diverse scientific practices and to move beyond the false binary of "Indigenous knowledge vs. science."

Particularly, as observed at the beginning of the table, there are many commonalities between the two perspectives, as listed below, and it is inappropriate to consider them as excessively different.

- Empirically based (sharing intellectual processes such as observing, looking for patterns, inferring, predicting, verifying, etc.).
- Elders and scientists exercise rational thoughts, and when doing so, they demonstrate communal characteristics (honesty, inquisitiveness, perseverance, and open-mindedness).
- Dynamic and historic, that is, change is based on past understandings.
- Outcomes are continually being revised in light of new observations and new ideas.
- Culture-based. Also, it is unnecessary to view one as superior to the other.
- Typically less explicitly place-based; by contrast, Indigenous ways of living in nature is often described as strongly place-based.

Although apparent differences exist in many aspects, these are fundamentally rooted in several ideas I have previously discussed, such as adopting a holistic perspective, harmonising human and natural elements, avoiding distinctions between subject and object, reconciling the ambiguity between artificial and natural entities, and recognising the impermanence of nature.

2.3.2 Responsibility for Environmental Issues in Discussion of Japanese Views of Nature

This section examines how responsibility for environmental issues has been discussed in relation to Japanese views of nature, contrasting these discussions with arguments commonly made about Western science. As I have consistently emphasised, it is imperative to exercise caution to avoid stereotyping when making comparisons with Western approaches.

In the Western context, Lynn White (1967) argueed that "by destroying pagan animism, Christianity made it possible to exploit nature in a mood of indifference to the feelings of natural objects." Watanabe (1991) also noted that "modern science and technology emerged from a view of nature rooted in the Judaeo-Christian tradition, and this viewpoint is thus largely responsible for the environmental destruction that modern science and technology have called forth," referencing Lynn White's viewpoint. This Western view on controlling nature is grounded in religions that consider humans as "having 'dominion over' other creatures," as exemplified in the Old Testament's "be fruitful, and multiply, and replenish the earth" (Suzuki, 2018). Importantly, these claims remain contested; the purpose here is to summarise a prominent critical interpretation rather than to treat it as the sole explanation.

This section focuses on the sense of responsibility concerning environmental issues within the context of Japanese views of nature; therefore, it will not further explore the Western conception of responsibility. Nonetheless, it is worth nothing that a distinct sense of accountability for environmental problems has developed in response to concerns about the societal and environmental impacts of science, technology and innovation, as reflected, for example, in the contemporary framework of Responsible Research and Innovation (RRI).

On the other hand, in Japanese views of nature, humans and nature are portrayed as harmonious and integrated, and the boundary between nature and artificiality is ambiguous. As Watanabe (1991) introduces, in the Fukuoka folk song *Tanko Bushi*, there is a phrase like "The chimneys are so tall! Poor moon! All enshrouded in smoke."

In this manner, from Japanese views of nature, there exists a propensity to consider all phenomena as *shizen*. Murakami (1977) describes a line of reasoning suggesting that if wastewater is permitted to remain unaltered, the soil will inherently filter it, or it will be processed within the sea as it is. This form of reasoning in Japan is sometimes discussed in connection with the Japanese phrase *mizu ni nagasu* (let it flow in the water). It is used to express the concept of disregarding past events as if they never happened and of refraining from assigning blame for what has transpired. Furthermore, it has been observed that the ambiguous distinction between the natural and the artificial may result in an inability to differentiate between nature itself and nature compromised by artificial intervention. Additionally, this ambiguity can lead to the inability to distinguish between natural disasters and those caused by human activity (Ito, 1999). With a sense of impermanence, Kamo no Chomei is often read as accepting natural disasters as fate and expressed an attitude that may be paraphrased as finding companionship in music and nature. If natural and man-made disasters cannot be distinguished, there may also be a possibility of inadvertently accepting man-made disasters as inevitable fate.

Watanabe (1991) argues that even when new scientific and technological developments harm the environment, people who make use of these advances often do not see themselves as responsible. This lack of accountability is rooted in a long-established view that humans and nature have historically existed on equal, mutually dependent terms. Because humans are not positioned as active agents and nature is not treated as an object, the idea that humans bear responsibility for nature does not always take shape.

Thus, it can be observed that Japanese views of nature may contribute to ambiguity or a reduced attribution of responsibility concerning environmental issues, as exemplified by the phrase *mizu ni nagasu*. However, the extent to which this idiom should be mapped onto environmental responsibility remains an empirical and interpretive question.

2.4 Towards a Better Integration of Japanese Views of Nature and Western Science, as Demonstrated by Kakishibu—Beyond *Wakon-Yosai*

Kakishibu featured in this publication as an instance of ecological innovation represents a traditional knowledge system developed through Japanese views of nature. Kakishibu is a fermented astringent persimmon juice and has been extensively utilised in Japan; it is often described as over a millennium. While more details about Kakishibu are in Chaps. 3 and 4, I will briefly highlight two main reasons for its widespread and lasting use.

The multifunctionality of Kakishibu is noteworthy. It is widely reported that when materials such as paper, wood, and cloth are coated or dyed with Kakishibu, they can acquire various properties including water repellency, enhanced durability, antibacterial and antiviral capabilities, as well as deodorising effects. Kakishibu can be seen as a traditional coating agent similar to plastic, although it might seem odd to describe Kakishibu as plastic, since Kakishibu has existed long before the development of petroleum-based plastics (e.g., Bakelite, 1907) (Kano, 2025).

The other aspect pertains to the sustainability of Kakishibu. Its production involves the use of persimmon fruit, which can be naturally harvested annually. The fermentation process in traditional production methods employs natural

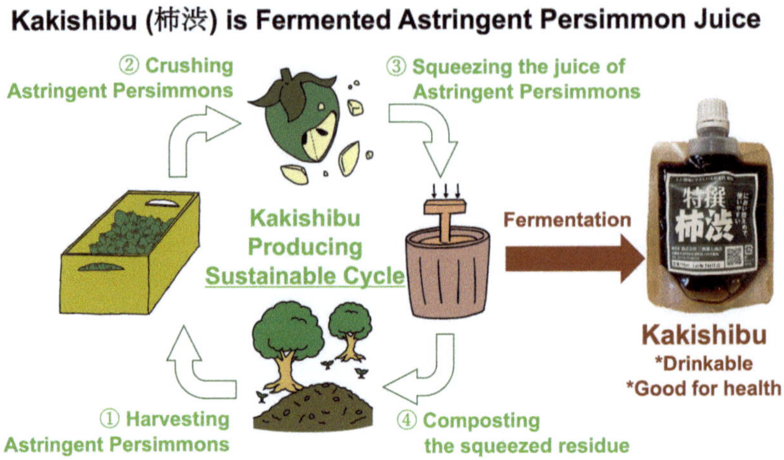

Fig. 2.2 The Kakishibu production cycle. (Translated into English and reprinted from Kano (2025))

fermentation facilitated by microorganisms present in the fruit. Kakishibu is generally regarded as biodegradable. Moreover, the extracted residue has been repurposed as fertiliser (Fig. 2.2). The primary constituent of Kakishibu is tannin, and in many contexts, tannins are extracted from various plant sources, sometimes involving the harvesting of bark or wood. In contrast, Kakishibu is characterised by a highly sustainable cycle of production and disposal.

Kakishibu exemplifies wisdom-in-action for survival, circular, and harmony with nature as a dynamic and multidimensional balancing of interrelationships for survival. It represents traditional knowledge that aligns with the typical Japanese ways of knowing nature.

Historically, the demand for Kakishibu significantly diminished following the large-scale introduction of petroleum-based plastic to Japan. This phenomenon may be referred to as reduction in Kakishibu due to the spread of plastic.

However, as concerns about the environmental impacts of plastics have intensified, there has been increasing reflection on the responsibilities of science, industry, and society. Consequently, efforts are being made to reduce plastic consumption. In this context, the widespread adoption of Kakishibu may facilitate a reduction in plastic usage and waste, signifying a paradigm shift away from the reduction in Kakishibu due to the spread of plastic phenomenon. This paradigm shift could also address the lament expressed by Torahiko Terada (1948), who mourned that "it is truly regrettable that in modern Japan, we are so engrossed in the enjoyment of natural blessings that we completely forget about avoiding natural disasters."

What is necessary to realise this? As previously discussed in Sect. 2.3.1, the current term nature in Japanese is often understood as encompassing both older meanings of *shizen* (e.g., as it is or spontaneity) and the European-originated notion of nature, introduced and institutionalised in the modern period. This amalgamation has led to some ambiguities within Japanese views of nature. However, as Yanabu (1977) observes that the moment has arrived to deliberately connect two traditions with distinct roots and, through that linkage, forge a renewed understanding of nature. In a similar spirit, Hideki Yukawa maintains that confronting environmental problems now requires bringing Western and Eastern philosophies into dialogue—a synthesis that joins deep currents of Eastern thought with Western physics grounded in modern rationalism. Furthermore, Poskett (2023) argues, in a global history of science, that the development of modern science has long relied on exchanges of ideas across different cultures, indicating that knowledge has always circulated internationally. Taken together, these accounts underscore that knowledge has never been confined within national or civilizational boundaries; it has long circulated internationally through continual interaction among cultures.

The specifics of the concepts necessary to achieve a better integration of Japanese views of nature with Western science will be examined in the concluding chapter, Chap. 9, but here I would like to present three perspectives.

The initial approach involves elucidating the components and mechanisms of Kakishibu through Western scientific methods, while concurrently respecting the Japanese way of knowing nature, as exemplified by Jokichi Takamine and Kikunae Ikeda in their work with *koji* and kelp. One of the authors of this publication, Tomoaki Matsuo, is a pioneer in this internationally recognised field; however, numerous questions remain unresolved. For additional information on Matsuo's endeavours, please refer to Chap. 4.

The second aspect involves cultivating a sense of responsibility towards environmental issues while contemplating the harmony between humans and nature. Drawing from the Japanese philosophical perspective on understanding nature, the phrase *mizu ni nagasu* (let it flow in the water) can be read as reflecting an ambiguous attribution of responsibility. Of course, this arises from perceiving humans and nature in harmony, despite the ambiguous boundary between artificial and natural elements. However, it is imperative to consciously delineate the distinction between artificial and natural components and develop a sense of accountability for environmental problems caused by artificial objects. Japan's per-capita plastic consumption is often reported as high, and plastic waste remains a major policy issue. Consequently, it is essential for Japan to assume greater responsibility regarding the plastic waste issue and to endeavour towards enhancing harmony between nature and humanity, thereby facilitating the goals of plastic reduction and disposability through the dissemination of Kakishibu.

Lastly, although it may appear contradictory to the initial point, I would like to emphasise the importance of respecting Japanese views of nature and refraining from excessive research and development or demands for Kakishibu. When advancing research related to Kakishibu, it is essential not to adopt an attitude of conquering nature. "Despite humans possessing knowledge, it is imperative to refrain from utilising it. While it is impossible to avoid acquiring knowledge, we must endeavour not to employ knowledge that poses a threat to humanity," as proposed by Joseph Needham, drawing upon Taoist philosophy (Nakayama et al., 1988), and Suzuki (2018), who referred to it as "the reservation of action." In the context of Kakishibu research and development, its functionality can be enhanced through excessive chemical modifications to its primary component, tannin, rendering it akin to plastic. Nonetheless, possessing the capability does not justify undertaking such measures. Caution is imperative. Examples of responsible research and innovation grounded in nature, informed by Japanese views of nature, can be found in both the Kakishibu factory case (Chap. 3) and the case of a global company (Chap. 7). I contend that respecting traditional methodologies, scaling up responsibly, and conducting fair trade at appropriate prices—characterised by Kakishibu factory—can forge a novel paradigm of responsible research and innovation rooted in the Japanese way of knowing nature, representing a synthesis of Eastern and Western perspectives.

The two kanji characters for *shizen*, 自然 can be taken to symbolise a meeting point between older meanings associated with *shizen* and the modern scientific concept of nature. In practice, Japanese usage has often moved between these senses without always making the distinction explicit. This unconscious differentiation has led to the emergence of peculiar terms such as *shizen* energy (renewable energgh), and has rendered responsibilities somewhat ambiguous.

However, Morinaga (1976) argues that meaningful contributions become possible only when we first recognize the differences between the East and the West, including their historical backgrounds. Building on this view, deliberately clarifying those distinctions—while also seeking deeper points of connection—may help generate new forms of innovation and strengthen a shared sense of responsibility. This represents a re-innovation rooted in previously established traditional knowledge, and it is our belief that this process of re-innovation could be framed as a form of "Responsible Research and Re-Innovation (RRRI)."

References

Aikenhead, G. S., & Ogawa, M. (2007). Indigenous knowledge and science revisited. *Cultural Studies of Science Education, 2*(3), 539–620. https://doi.org/10.1007/s11422-007-9067-8

de Waal, F. B. (2003). Silent invasion: Imanishi's primatology and cultural bias in science. *Animal Cognition, 6*(4), 293–299. https://doi.org/10.1007/s10071-003-0197-4

Fuke, N. (2008). "Umami" o kaimei shita Ikeda Kikunae: Riron kagaku to ōyō kagaku o jizai ni ōrai (Kikunae Ikeda, who elucidated "umami": Freely moving between theoretical chemistry and applied chemistry). *Kindai Nihon no Sosetsushi (History of Modern Japan's Founding), 5*, 3–13.

Fukuzawa, Y. (1866). *Seiyo jijo* (Conditions in the West). Retrieved March 24, 2025, from https://dcollections.lib.keio.ac.jp/ja/fukuzawa/a02/3

Ito, S. (1995). Yukawa Hideki no shizenkan (Yukawa Hideki's view of nature). In S. Ito (Ed.), *Nihonjin no shizenkan (Japanese views of nature)* (pp. 463–478). Kawade Shobo Shinsha.

Ito, S. (1999). *Shizen (Nature)*. Sanseido.

Kano, K. (2025). Nihon no dento chi "kakishibu" ni yoru datsu purasuchikku e—Riyo kachi kainyu wo toriireta STEAM kyoiku puroguramu dezain—(moving away from plastic with traditional Japanese knowledge of Kakishibu: STEAM education program design incorporating value intervention). In M. Taniguchi (Ed.), *Ninshiki to kodo no chikyuu kankyogaku (global environmental studies of perception and behavior)*. Kokonshoin.

Kono, T. (2020). Ikeda Kikunae: Umami no hakken – Rigaku to Oyo no sokoku no nakade (Kikunae Ikeda: The discovery of umami – The struggle between theoretical and applied science). *Chemical Education, 68*(11), 474–477.

Matsuzawa, T. (2009). Reichoruigaku 60-nen to Imanishi Kinji: Sekai no reichoruigaku ni okeru Nihon no koken (sixty years of primatology and Kinji Imanishi: Japan's contribution to world primatology). *Primate Research, 24*(3), 187–196.

Morinaga, H. (1976). Nihonjin ni mo kagaku ga dekiru ka? (Can Japanese do science?). *Shizen (Nature), 31*(1), 52–58.
Murakami, Y. (1977). *Kagaku, tetsugaku, shinko (science, philosophy, faith)*. Third Civilization Society.
Nakayama, S., Matsumoto, S., & Ushiyama, K. (1988). Josefu Nidamu no sekai *(the world of Joseph Needham)*. Nihon Chiiki Shakai Kenkyujo.
Needham, J., & Ling, W. (1956). *Science and civilisation in China (Vol. 2): History of scientific thought*. Cambridge University Press.
Nelson, G., Chandrashekar, J., Hoon, M. A., Feng, L., Zhao, G., Ryba, N. J., & Zuker, C. S. (2002). An amino-acid taste receptor. *Nature, 416*(6877), 199–202. https://doi.org/10.1038/nature726
Institute for Gender Studies at Ochanomizu University. (2003). *Michiyo Tsujimura material catalog*. Retrieved March 24, 2025, from https://teapot.lib.ocha.ac.jp/records/4091
OECD. (2022). *Education at a glance 2021*. Retrieved March 24, 2025, from https://www.oecd.org/en/publications/education-at-a-glance-2021_b35a14e5-en.html
Poskett, J. (2023). *Horizons: A global history of science*. Penguin.
Cai, M. (1992). Yukawa Hideki no soryushi riron kenkyu to chugoku no roso tetsugaku (Hideki Yukawa's particle theory research and Chinese Laozi philosophy). *Soryushiron Kenkyu (Particle Theory Research), 85*(2), 114–125.
Sessyu. (1474). *Shihombokugasansuizu*. Retrieved August 10, 2025, from https://bunka.nii.ac.jp/heritages/detail/150691
Suzuki, D. T. (1960). Lectures on Zen Buddhism. In D. T. Suzuki, E. Fromm, & R. De Martino (Eds.), *Zen Buddhism and psychoanalysis* (pp. 1–76). George Allen & Unwin.
Suzuki, S. (2018). *Nihonjin no shizenkan* (Japanese views of nature).
Suzuki, J. (2022). *Takamine Jokichi bunshu: Ikani shite hatsumei kokumin to narubeki ka (Jokichi Takamine collection: How to become an inventive nation)*. Iwanami Shoten.
Terada, T. (1948). Nihonjin no shizenkan (Japanese views of nature). In T. Komiya (Ed.), *Torahiko Terada essay collection*.
Watanabe, M. (1976). *Nihonjin to kindai kagaku: Seiyo e no taio to kadai (The Japanese and Western science: Response and challenges to the West)*. Iwanami Shoten.
Watanabe, M. (1991). *The Japanese and Western science* (O. T. Benfey, Trans.; Foreword by E. O. Reischauer). University of Pennsylvania Press. (Original work published 1976)
Watanabe, M. (1995). Kindai ni okeru Nihonjin no shizenkan – Seiyo tono hikaku ni oite (Japanese views of nature in modern times: In comparison with the West). In S. Ito (Ed.), *Nihonjin no shizenkan (Japanese views of nature)* (pp. 329–370). Kawade Shobo Shinsha.
Watanabe, M. (2000). *Bunka toshiteno kindai kagaku: Rekishiteki gakusaiteki shiten kara (Modern science as a culture: From a historical and interdisciplinary perspective)*. Kodansha.
White, L. (1967). The historical roots of our ecologic crisis. *Science, 155*(3767), 1203–1207.
Yamagiwa, J. (2009). Nihon no reichoruigaku: Rekishi to tenbo (primatology in Japan: History and prospects). *Primate Research, 24*(3), 183–186.
Yanabu, A. (1977). *Honyaku no shiso "shizen" to NATURE (Translation thought: "Nature" and NATURE)*. Heibonsha.

Kei Kano, PhD in life sciences, is a professor at the Science Communication Laboratory in the Faculty of Education at Shiga University, Japan. He is a visiting fellow at the Australian National Centre for the Public Awareness of Science at the Australian National University. He also serves as the chair of foreign affairs at the Young Academy of Japan, the Science Council of Japan, and an associate member of the Science Council of Japan. He has won the Prize for Science and Technology from the Ministry of Education, Culture, Sports, Science and Technology (MEXT) in Japan twice, in 2014 and 2017, for his work in science communication practices and research.

Part II
History and Science of Kakishibu

Chapter 3
History of Kakishibu—Before and After the Plastic Era

Tomokazu Goto, Yoshitsugu Mimasu, and Takeo Mimasu

Abstract Kakishibu is made by crushing unripe astringent persimmons, squeezing them to extract the juice, and then fermenting and aging the juice for a long period of time. This liquid, which contains large amounts of persimmon tannin, has been used for various purposes since ancient times and has been deeply involved in everyday life and industry. Among these uses, Kakishibu has been employed as a paint or dye to impart water-resistant, preservative, insect-repellent, strength, and durability properties to natural materials such as wood, cloth and "*washi*," traditional Japanese hand-made paper selected as Intangible Cultural Heritage by UNESCO. It has been indispensable in everyday items, wooden architecture, and the production of cloth bags used at the pressing stage in the brewing industry. However, with the advent of the plastic age, demand for natural materials declined significantly, and demand for Kakishibu also plummeted. Amidst this situation, Kakishibu came to be widely used as a fining agent for sake in the sake brewing industry, and demand for it was revived. Furthermore, it began to be used in various other applications, such as health foods, processed foods, and cosmetics. Kakishibu is environmentally friendly, safe for humans, and has various excellent properties, which has led to its widespread use. Due to its versatility, Kakishibu has been used in many different ways in response to the changing times and demands of society and continues to be used today. This chapter outlines the history of Kakishibu prior to and following the era of the plastic society, as well as the future prospects of Kakishibu.

Keywords Kakishibu · Persimmon tannin · Paint · Dye · Fining agent · Food additive · Environmentally friendly · Safe for the human body · Multifunctionality · Multipurpose

---

T. Goto (✉) · Y. Mimasu · T. Mimasu
Mimasu Kashichi Shouten Co., Ltd., Kyoto, Japan
e-mail: kaki@mimasu.net; kaki@mimasu.net

K. Kano et al. (eds.), *Traditional Knowledge for Green Innovation with Japanese Persimmon Tannin, Kakishibu*, https://doi.org/10.1007/978-981-95-7805-4_3

3.1 What Is "Kakishibu"?

Kakishibu is a product made by fermenting and aging juice squeezed from unripe, astringent persimmons, with persimmon tannin as its main component. It has been used for various purposes thanks to the collective wisdom of experience over time and has been regularly used in daily life. In particular, it has been used as a coating and dye for natural materials such as wood, bamboo, "*washi*" (traditional Japanese hand-made paper selected as Intangible Cultural Heritage by UNESCO), cloth, and thread, with the aim of making these materials water-resistant, preservative, insect-proof, strengthened, and durable.

However, with the start of plastic production in the late Showa 20s (the early 1950s) and its rapid spread thereafter (Tajima, 2016), the demand for natural materials greatly decreased, along with the demand for Kakishibu. Amid such circumstances, in the Showa 40s (the decade from 1965 to 1974), Kakishibu began to be widely used instead as a fining agent for sake in the brewing industry, leading to a renewed surge in demand for Kakishibu (Conference on the Specifications of Sake Treating Materials, 1998). On the other hand, in an era dominated by the convenience of the chemical industry, pollution eroded people's health and the natural environment. Amid these serious negative effects, the momentum for health consciousness and consideration for the environment increased, and natural materials were re-evaluated. The use of Kakishibu, which had a small demand except for the use of fining agents, gradually increased for purposes such as coating and dyeing. Furthermore, the many excellent functionalities of Kakishibu began to attract attention, and it came to be used in various fields.

Kakishibu has a long history that has blended into people's lives. This history is filled with the rich usage experience of Japanese ancestors. Here, we would like to explore a part of the over 1000-year history of Kakishibu in Japan, along with the humble company history of "Mimasu Kashichi Shouten Co., Ltd.," who has made Kakishibu production their livelihood.

3.1.1 The Raw Materials of Kakishibu

Persimmons are broadly divided into sweet and astringent types and are classified into the following four types based on the presence or absence of seeds, sweetness and astringency (Kometani & Takemori, 2016; Kanzaki, 2016; Kikuchi, 1948):

1. Complete Sweet Persimmon (Pollination Constant Non-astringent): A variety that loses its astringency on the tree, regardless of the presence of seeds.
2. Incomplete Sweet Persimmon (Pollination Variant Non-astringent): A variety that completely loses its astringency on the tree when it has many seeds but does not lose its astringency at all when it has no seeds.
3. Incomplete Astringent Persimmon (Pollination Variant Astringent): A variety that loses its astringency only around the area where seeds are formed.

Fig. 3.1 Astringent persimmon fruit (raw material for Kakishibu)

4. Complete Astringent Persimmon (Pollination Constant Astringent): A variety that does not completely lose its astringency on the tree, regardless of the presence of seeds. It exhibits astringency except when fully ripe.

Among these, the "Complete Astringent Persimmon" variety, which contains more persimmon tannin compared to the other three classifications, is optimal as a raw material for Kakishibu (Fig. 3.1). The harvesting period for the unripe fruit used as raw material is from around July to September. Harvesting is done at the time when the persimmon tannin content is highest, but this time varies depending on the persimmon variety.

3.1.2 Manufacturing Method

The general manufacturing method for Kakishibu is shown below (Fig. 3.2). Fresh persimmon fruit is used as the raw material, which is first washed and crushed, then juiced. The resulting juice becomes Kakishibu after fermentation and storage/aging for about one to three years (Figs. 3.2 and 3.3). Powdered Kakishibu is manufactured by drying Kakishibu. Granulated Kakishibu is manufactured by granulating powdered Kakishibu. The residue produced during the manufacture of Kakishibu is effectively used as, for example, a soil conditioner.

Notably, the methods of making Kakishibu and red wine, which are ancient brews, are almost the same as each follows a similar process (Commentary on Standard Analytical Methods Committee, 2017; National Research Institute of Brewing, 2018).

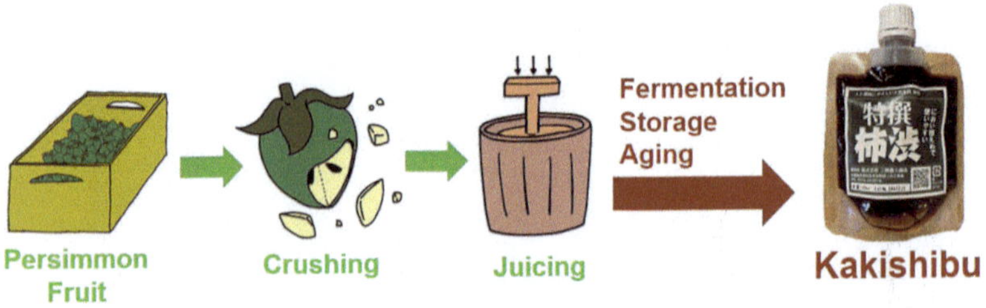

Fig. 3.2 Process of Kakishibu production

Fig. 3.3 Storage and maturation of Kakishibu

3.1.3 Quality Indicators

In the early modern period, the quality of Kakishibu was judged solely by experience. However, from around 1879 or 1880, the concentration of Kakishibu began to be expressed in a measurement called Baumé degree, which was used to judge its quality (Imai, 2003). Baumé degree is a unit of specific gravity, and the higher the Baumé degree, the higher the concentration of Kakishibu (Imai, 2003; Nakabayashi, 1971), and the higher price it fetched. The appropriate Baumé degree varied depending on the application, with 5–6 degrees for fining agents, 3–4 degrees for paints, and 2 degrees for fishing nets (Imai, 2003). However, there are exceptions. For example, if the sugar content in the Kakishibu was high, the Baumé degree would also be high, so the concentration of Kakishibu and Baumé degree did not always correlate (Nakabayashi, 1971).

Therefore, the National Tax Agency established "*rikika*," the titer of the colloidal titration method, as a more reliable quality indicator than Baumé degree for Kakishibu used as a fining agent in alcoholic beverages such as sake and mirin. The rikika of Kakishibu is correlated to the amount of persimmon tannin in the Kakishibu. The National Tax Agency has established specifications and test methods for Kakishibu used as a fining agent for alcoholic beverages, and has set standards for rikika, fining effect, lead, arsenic, and hiochi bacteria (National Tax Agency, 1961, 1997). Importantly, unlike Kakishibu used as a fining agent for alcoholic beverages, there are currently no common quality indicators or industry standards for Kakishibu used for paints and dyes.

3.1.4 Multifunctionality

Kakishibu has various functions, the main ones of which are:

1. *Water-resistance, preservative, toughness, durability*: When Kakishibu dries, the main component, persimmon tannin, oxidizes and polymerizes to form an insoluble film, giving the material water-resistant, preservative, toughness, and durability properties. It has been used for coating wood and *washi*, Japanese paper, and for dyeing fabrics and threads made of hemp, cotton, and other materials.
2. *Insect Repellent*: Kakishibu is used as an insect repellent for clothing and *washi* (Fujiki, 1940; Tanaka, 1991; Imai, 2003).
3. *Adhesive*: Kakishibu is used to bond *washi* to *washi* (Imai, 2003). It is also mixed with glue, which is made of a half-and-half mixture of tapioca and bracken powder, to strengthen its adhesive power (Imai, 2003).
4. *Color Development*: Wood and *washi* coated with Kakishibu, and fabrics and threads dyed with Kakishibu, develop a unique brown color over time, and this color has fascinated people since ancient times.
5. *Aggregation with Proteins*: Kakishibu has strong aggregative action with proteins, forming complexes (Japan Food Additives Association, 2021; Conference on the Specifications of Sake Treating Materials, 1998).
6. *Binding with Bases*: Kakishibu shows a strong affinity for basic compounds such as alkaloids, forming complexes (Japan Food Additives Association, 2021).
7. *Complex Formation with Metal Ions*: Kakishibu shows a strong affinity for ions of heavy metals such as iron, copper, chromium, gold, uranium and cesium, forming complexes (Japan Food Additives Association, 2021; Sakaguchi, 1996; Oto et al., 2012; Inoue, 2007, 2014).
8. *Reaction with Aldehydes*: Kakishibu reacts with various aldehydes such as formaldehyde, acetaldehyde, and glutaraldehyde to form a polymerized gel (Sakaguchi, 1996; Tanaka, 1997).
9. *Antioxidant*: Kakishibu has a strong antioxidant effect. Comparative data on the antioxidant capacity of our powdered Kakishibu (product name: Asutorin P) and various foods is presented in Tables 3.1 and 3.2.

Table 3.1 Antioxidant capacity of powdered Kakishibu (Asutorin P)

Sample	Superoxide scavenging capacity[a] (SOD units/g)	DPPH radical scavenging capacity[b] (DPPH units/g)
Kakishibu Powder (Asutorin P)	890,000	6,400

Tested by Japan Food Research Laboratories at the authors' request

[a]Electron Spin Resonance (ESR) method. The superoxide scavenging capacity equivalent to 1 unit of SOD is expressed as "1 SOD unit." One SOD unit refers to the unit defined by J. M. McCord and I. Fridovich (1969)

[b]50% ethanol was used as the extraction solvent. Visible absorption method. DPPH radical scavenging capacity equivalent to 1μmol of Trolox is expressed as "1 DPPH unit," according to Suda (2000)

Table 3.2 Antioxidant capacity of various foods

Sample	Superoxide scavenging capacity[a] (SOD units/g)	DPPH radical scavenging capacity[b] (DPPH units/g)
Ascorbic acid	330,000	5,400
Black tea	75,000	1,700
Hojicha (roasted green tea)	70,000	1,300
Instant coffee	26,000	650
Rooibos tea	3,900	330
Cocoa	1,800	170
Azuki beans	1,800	12
Soybeans	1,600	3
Chili pepper	1,200	25
Grape juice (purple)	800	11
Bell pepper (red)	760	10
Perilla (green)	510	16
Ginger	230	11
Lemon	210	5
Orange juice (100%)	110	3
Black sesame	60	5

Source (modified with permission): Japan Food Research Laboratories (2003)

[a]Electron Spin Resonance (ESR) method. The superoxide scavenging capacity equivalent to 1 unit of SOD is expressed as "1 SOD unit." One SOD unit refers to the unit defined by J. M. McCord and I. Fridovich (1969)

[b]50% ethanol was used as the extraction solvent. Visible absorption method. DPPH radical scavenging capacity equivalent to 1 μmol of Trolox is expressed as "1 DPPH unit," according to Suda (2000)

10. *Antibacterial*: Kakishibu prevents the growth of miscellaneous bacteria (Ogata, 1986). It also has antibacterial properties against *Staphylococcus aureus* (Details will be described later in Sect. 3.2.7), *E. coli* and MRSA (Methicillin-resistant *Staphylococcus aureus*) (information from symposium presentation by Yasuko Kobayashi in 2018).

11. *Antiviral*: Kakishibu shows antiviral activity against non-enveloped viruses such as feline calicivirus, murine norovirus, poliovirus, coxsackievirus, adenovirus, rotavirus, and enveloped viruses such as human influenza virus, avian influenza virus, herpes simplex virus type 1, vesicular stomatitis virus, Sendai virus, and Newcastle disease virus (Ueda et al., 2013).
12. *Deodorant*: Tests with Kakishibu-dyed cotton cloth have shown strong deodorizing effects. Details will be described later in Sect. 3.2.7.
13. *Odor prevention with anti-bacterial*: Tests with Kakishibu-dyed cotton cloth have shown odor prevention benefits associated with anti-bacterial effects. Details will be described later in Sect. 3.2.7.
14. *Contact cooling sensation*: Kakishibu-dyed cotton cloth provides a refreshing sensation throughout the year, especially a cooling sensation in the summer, by which people can stand the humid summer in Japan. Details will be described later in Sect. 3.2.7.

3.1.5 Multipurpose

Kakishibu has been used for various purposes. It has been used as a natural paint and dye for everyday items and wooden buildings, including temples and shrines. Kakishibu is also based on traditional knowledge. Traditional knowledge is often passed on through oral tradition, and there is a tendency for few documents to be passed down (Aikenhead & Ogawa, 2007). In this context, the efforts of Imai (2003), who collected and organized a wide range of documents and conducted interview surveys to compile a wide range of traditional knowledge about "Kakishibu," are particularly valuable and important. The main industrial uses of Kakishibu, such as in the brewing industry for *sakabukuro*, cloth bags used at pressing stage of sake, fining agents, fishing nets, and food and cosmetics, will be discussed later in Sect. 3.2. Here, we want to introduce the daily use of Kakishibu on wooden household items, *washi*, cloth, threads, and in wooden architecture.

3.1.5.1 Wooden Household Items

The public used to apply Kakishibu to almost all wooden containers that came into contact with water in their daily lives, such as bowls, water buckets, mangers, ladles, tubs, bath buckets, and lunch boxes. The role of wooden containers in daily life was significant and was not until the end of the Taisho era (1912–1926) that a transition to pottery started. Persimmons were cultivated in the gardens of ordinary households.

3.1.5.2 *Washi,* Japanese Paper

One important application of Kakishibu revolved around *washi*, Japanese paper, which has been used in the daily lives of Japanese people since ancient times. When *washi* was frequently used in daily life, Kakishibu was applied to *washi* to provide water-resistant, preservative properties and to increase its toughness. Daily necessities made of *washi* coated with Kakishibu include fans, wrapping paper, mats, and clothing. The details about them are provided below.

1. *Shibu-uchiwa,* fans made from *washi* coated with Kakishibu, were used in kitchens and other areas to blow air to flames. They were durable and long-lasting and thus were in high demand in the early modern period. They were actively produced until just after the end of World War II.
2. *Shibugami, washi* layered and coated with Kakishibu, were used as wrapping paper. Its use was quite widespread from the late medieval period to the beginning of the early modern period. Various items were wrapped in *shibugami* and transported to distant places.
3. *Shibugami* was also used as a mat for the purpose of dust removal and insect repellent in the late Edo period (late eighteenth century—1867). *Shibugami* mats have been used until the Showa era (1926–1989), and they remained the longest and had the highest utility value in the history of *shibugami* use.
4. *Kamiko,* clothing made by gluing *washi* together, coating it with Kakishibu, drying it, kneading it well, and then beating it with a wooden mallet. From the middle of the Edo period (late seventeenth century–late eighteenth century), when cheaper *washi* became available, *kamiko* became popular among the public because it was warm, light, foldable, not bulky, and affordable.

3.1.5.3 Cloth and Threads

Kakishibu was used for dyeing cloth and thread because of its water-resistant and preservative properties. The practice of dyeing clothes with Kakishibu can be seen in documents from the early Edo period (1603–late seventeenth century) (Imai, 2003). A book published in the mid-eighteenth century details the method of dyeing cloth with Kakishibu, which was done regularly in each household in the middle of the Edo period (late seventeenth century–late eighteenth century) (Imai, 2003).

3.1.5.4 Wooden Architecture

Kakishibu was used as a building paint because of its water-resistant and preservative properties from the Edo period (1603–1867) to around the Taisho era (1912–1926). It was applied to the board fences and corridors of shrines and temples, the wooden walls of castles, and so on. In addition to shrines and temples, in the houses of the late Edo period (late eighteenth century—1867), it was also common to paint the wooden walls and fences with Kakishibu. Even after the Meiji era (1868–1912), wooden elementary school buildings and local barracks were painted with Kakishibu. Later, when Western-style houses used paint, Kakishibu was used as the main paint in ordinary houses at that time in Japan. In addition, Kakishibu was used on the interior building materials of wooden buildings—pillars, lintels, kitchens, baths, toilets—with the aim of making the wood last longer due to its preservative and insect repellent effects. However, with the advent of creosote and other chemical preservatives and paints in the Meiji era, Kakishibu quickly disappeared as a paint.

3.1.6 Environmentally Friendly and Safe for Humans

Kakishibu is made from persimmon fruit and is manufactured by combining physical methods such as crushing and juicing, biological action through fermentation, and storage and maturation processes, without any chemical synthesis process. There is no literature on the biodegradability of Kakishibu, but it is clear from its manufacturing process that Kakishibu is composed of 100% natural materials and from the fact that Kakishibu has not caused any harm to the environment in its long history of use. Its biodegradability is clear. For example, the residue after the production of Kakishibu, which still contains a lot of persimmon tannin, makes a good soil conditioner. Our Kakishibu company, Mimasu Kashichi Shouten Co., Ltd., uses the residue in our own persimmon orchard. As a side note, around May, many beetle larvae can be seen in the area where the residue has been spread in the persimmon orchard, creating a comfortable natural habitat for beetle larvae. This is a natural process that starts with persimmons. Furthermore, Kofujita (2013) reported on the biodegradation mechanism of condensed tannins, including persimmon tannin. In this way, Kakishibu is an environmentally friendly material that harmonizes with the environment.

Additionally, Kakishibu has been used as a traditional folk medicine. Here are some main uses and benefits:

1. Stroke and hypertension: Kakishibu is a unique Japanese folk medicine that has been used for stroke and hypertension since ancient times. The "Secrets of Practical Nursing at Home" (commonly known as the "Red Book" (Tsukita, 1984)) has been highly regarded and widely disseminated nationwide as an encyclopedia of folk remedies, includes a section stating, "Drinking Kakishibu cures stroke," and details the method of administration (Tsukita, 1984, p. 855; Imai, 2003). The effect of Kakishibu on hypertension has been demonstrated in pharmacological studies using rabbits, showing that Kakishibu has a blood pressure-lowering effect, and in clinical trials showing a hypotensive effect on hypertensive patients (Ogata, 1986).
2. Burns: Kakishibu has been said to be effective when applied to burns. The "Red Book" also lists it as a "specific medicine for burns" (Tsukita, 1984, p. 483;

Imai, 2003). This is thought to be due to the adsorption of persimmon tannin in human tissue and its astringent effect, which constricts and suppresses cell swelling, preventing the skin from reddening and swelling due to high heat and preventing blister formation (Ogata, 1986). The authors' experience using Kakishibu as a burn treatment also supports its effectiveness.

3. Frostbite and chaps: It has been said since ancient times that applying Kakishibu to frostbite makes it heal easily. The "Red Book" also lists the application of Kakishibu as one of the "folk remedies that miraculously cure frostbite" (Tsukita, 1984, p. 439; Imai, 2003). This is thought to be because, like its effectiveness on burns, Kakishibu is well adsorbed to tissue, protecting and tightening it through its astringent effect, and suppresses bacterial proliferation (Ogata, 1986). Kakishibu has also been used for chaps. The authors use it in their daily lives and have experienced its effectiveness.
4. Hangovers: Kakishibu has been known and used since ancient times for its effectiveness against hangovers. There are reports on the mechanism of its effect by Ogata (1976, 1986) and Itamura (2011), among others.
5. Snake venom: There are local tales about how to deal with bites from venomous snakes like the Mamushi using Kakishibu (Imai, 2003). The "Red Book" introduces a method using Kakishibu in the section on "First Aid for Snake Bites" (Tsukita, 1984, p. 493; Imai, 2003). Later, it was reported multiple times that Kakishibu has a detoxifying effect against snake venom (Ogata, 1986; Imai, 2003).
6. Others: Kakishibu has been known since ancient times to be effective against diarrhea, bruises, hiccups, etc. (Ogata, 1986).

Thus, Kakishibu has been used by people as a folk medicine since ancient times.

3.2 History of Mimasu Kashichi Shouten: Expansion and Portfolio Change Due to Multifunctionality

Section 3.1 outlined the general properties of Kakishibu. Here, we will examine how its applications have evolved over time. Notably, the range of Kakishibu uses experienced significant changes before and after the advent of plastics. Specifically, there was a shift from coating and dyeing applications to additive uses in sake and food products. Focusing on our company's history spanning over 150 years, we will now describe how Kakishibu applications expanded and how its portfolio developed. Our company, "Mimasu Kashichi Shouten Co., Ltd.," manufactures and sells Kakishibu in the Kansai Science City (Keihanna Science City) area, which spans the hills of Kyoto, Osaka, and Nara. The company was founded in the Meiji 5 (1872). We have been making Kakishibu and supporting traditional industries in this area long before the birth of the region's science city, where cutting-edge research and technology development are advancing. The Yamashiro region, where the science city is located, is one of the erstwhile three major production areas of

Kakishibu, along with Hiroshima and Gifu, and although it does not reach the production volumes of the past, it is still a nationwide production area for Kakishibu.

Nowadays, the number of "*shibuya*," traditional small-scale Kakishibu manufacturing and sales companies, in this region has decreased, but in the past, there were shibuya in every village. Kakishibu was transported by boat to wholesalers in Osaka using the waterways of the Kizugawa River. This was the era when the company's founder, Takejiro Mimasu, started the Kakishibu manufacturing business. Takejiro's wife, Taki, came from a family that had been running a *shibuya* for generations in the nearby area, and after their marriage, the couple helped Taki's family to press persimmons to make Kakishibu.

At one point, Takejiro asked if he could have some of the pressed residue to press again and make more Kakishibu. They agreed, saying, "You can take it home and press it." So, they started pressing the residue at their place. This was the beginning of their business. Later, Takejiro's eldest son, Kashichi Mimasu, took over the business, and Kashichi's eldest son, Takeo Mimasu, one of this chapter's co-authors, established Mimasu Kashichi Shouten Co., Ltd. in 1968 and became its President. In 2003, Takeo became chairman, and his eldest son, Yoshitsugu Mimasu, another co-author, became President, a position he holds to this day.

3.2.1 Applications for Coating and Dyeing

Until around the early Showa 30s (1955–1959), most of our Kakishibu was used for coating and dyeing purposes, mainly for water-resistance, preservation, and reinforcement of materials. It was primarily used to dye *sakabukuro*, cloth bags used at the pressing stage of making sake and soy sauce, and the filtering cloth for soy sauce, for which there was a high demand. By dyeing the pressing bags and filtering cloth with Kakishibu, their strength and durability were improved, and the ease of separating the lees from the pressing bags and filtering cloth was also improved, making Kakishibu dyeing a very important process. It was also used as a paint for sake brewing tools such as wooden tubs, barrels, and casks. At that time, Kakishibu was used as a fining agent for sake only when the sake was very cloudy.

Moreover, dyeing fishing nets with Kakishibu enhanced their durability and longevity, while also facilitating improved handling by preventing tangling. Consequently, Kakishibu continued to be employed for dyeing fishing nets until approximately the Meiji period (1868–1912) (Imai, 2003). According to the limited oral traditions preserved within our company, it had not been used since long before World War II. It is likely that it was still used for fishing nets when our company was in its early years (1872–). Kakishibu was also widely used for dyeing "*nori* nets," nets for seaweed farming, and our company sold them to various fisheries cooperatives under the product name "Noriamin." During the Edo period, seaweed farming appears to have begun in the Enpo era (1673–1681) (Fujita, 2002). Nets began to be used in the Taisho era (1912–1926), and Manila hemp, wisteria vines, among other types, were used. From the late Taisho to the early Showa period (the Showa period:

1926–1989), palm and coir yarn began to be used, and from the Showa 20s (1945–1954) to the early 30s (the late 1950s), coconut nets were used for seaweed farming (Miyagi Prefecture Fisheries Cooperative Association's website. https://www.jf-miyagi.com/omotenashi/html/zukan14b.html; Kawamura, 2006).

Kakishibu was used to dye these natural fibers. In seaweed farming, diatoms that attach to the seaweed nets are a problem as harmful algae, and when they attach to the seaweed nets, they inhibit the growth of seaweed buds and the attachment of monospores (Ohgai, 1986). When the seaweed nets were dyed with Kakishibu, diatoms were less likely to attach to the nets, and the seaweed grew well. In addition, there were advantages such as the seaweed nets becoming stronger and lasting longer, and the handling of the nets improving (e.g., the nets not getting tangled) when dyed with Kakishibu.

However, with the rapid spread of plastics from the late Showa 20s (the early 1950s) onwards (Tajima, 2016), the production of chemical fiber-made *sakabukuro*, such as vinylon, nylon, and tetrone, began around 1960 (Kimishima & Fujii, 1966; Kida, 1963). Their use quickly progressed (Kida, 1963), and by around 1965, they were widely put into practical use (Kimishima & Fujii, 1966; Kimura, 1966). As a result, the use of cotton sake bags drastically decreased (Kimishima & Fujii, 1966), and Kakishibu was no longer used. Wooden sake brewing tools such as tubs, barrels, and casks also disappeared (Arakawa, 1964), and Kakishibu use similarly declined. In the soy sauce industry, the development of synthetic fiber filtering cloths made of vinylon, nylon, and tetrone advanced from around 1955 and rapidly spread to companies throughout the industry (Tochikura, 1994; Oguri, 2008). Consequently, cotton filtering cloths disappeared, and the use of Kakishibu also ceased (Tochikura, 1994).

Regarding seaweed nets, the use of vinylon seaweed nets was introduced in 1958 in the Ariake Sea, Saga Prefecture, a key area for seaweed farming (Kawamura, 2006), and vinylon seaweed nets were fully adopted in seaweed farming in Ehime Prefecture by 1959 (Ehime Prefectural History Compilation Committee, Social and Economic Affairs Subcommittee I, 1985), although synthetic fibers like nylon and vinylon started appearing from the mid-Showa 20s (early 1950s) (Kawamura, 2006). Therefore, it seems that natural fiber seaweed nets were used at least until around the mid-Showa 30s (the late 1950s), and Kakishibu was also used at least until then. Even with synthetic fiber seaweed nets, if they were dyed with Kakishibu and then thoroughly dried, Kakishibu would not easily come off, so Kakishibu was also used for synthetic fiber seaweed nets. However, Kakishibu eventually ended its role as a dye for seaweed nets. Many in the industry went out of business, and our company also faced a difficult situation. After this, the demand for coating and dyeing applications never recovered to its former level, but a small demand continued.

However, in recent years, as the trend towards health consciousness, sustainability, and environmental friendliness has grown, natural materials have been reevaluated, and attention has once again been focused on Kakishibu as a coating and dye. Furthermore, various excellent functionalities of Kakishibu have been discovered, and the range of uses for Kakishibu expanded. For example, our product "Mimasu Granule Kakishibu," a granulated form of liquid coating and dyeing Kakishibu, has excellent preservation, transportability, and convenience and is increasing in demand not only domestically but also overseas (details will be described later in Sect. 3.2.8).

3.2.2 Use in Sake—As a Fining Agent

As the use of Kakishibu for coating and dyeing declined with the introduction of plastic, its use as a fining agent for sake increased. Here, we explain in detail the use of Kakishibu as a fining agent for sake.

Kakishibu as a fining agent is used in alcoholic beverages such as sake and mirin (sweet sake) and in liquid foods such as soy sauce, vinegar, and honey. Here, we will mainly discuss on its use in sake (Conference on the Specifications of Sake Treating Materials, 1998).

The use of Kakishibu as a fining agent in sake seems to have started around the middle of the Edo period (late seventeenth century–late eighteenth century), and it began to be used in earnest from the middle of the Meiji period (1868–1912) (Kuruma, 1975). There are no known sources of old literature that record the use of Kakishibu as a fining agent, but among sake brewers and brewery owners, the addition of Kakishibu has been passed down over time as a customary technique for remedying protein-turbid sake.

After the post-World War II recovery period, sake consumption increased significantly during the high economic growth period of the Showa 30s (1955–1964). In this context, by the late Showa 30s (the early 1960s), the clarity or transparency of bottled sake began to attract attention due to changes in people's preferences and demand for clear sake increased in the market. Against this backdrop, more research was conducted on Kakishibu as a fining agent for sake (Akiyama et al., 1969; Mena & Nunokawa, 1971), and by the Showa 40s (1965–1974), fining using Kakishibu became widespread. The use of Kakishibu by sake manufacturers shifted from a dye for sake bags to a fining agent, and Kakishibu began to be used again, allowing Kakishibu manufacturers, who were in a difficult business situation, to recover.

We will now briefly explain the fining mechanism of sake using Kakishibu (Conference on the Specifications of Sake Treating Materials, 1998; Commentary on Standard Analytical Methods Committee, 2017; Brewing Society of Japan, 1993), and then briefly introduce the history of our company's development of fining agents.

Sake can sometimes become cloudy, which can reduce its commercial value. The cloudiness in sake is primarily due to the action of hiochi bacteria and protein turbidity. The latter occurs when enzymes remaining in the sake become insoluble microparticles due to heat denaturation during pasteurization, and these microparticles grow to a visible size during storage. In sake, protein turbidity can sometimes occur by happens, but it usually does not accompany changes in aroma or flavor. By fining the sake before shipment, it can be finished as a clear and lustrous (highly transparent) sake.

Fining in the context of alcoholic beverages refers to the process of removing turbid substances and the causative substances of turbidity in the beverage, thereby improving the transparency of the beverage and preventing the occurrence of turbidity (National Tax Agency, 1997). There are two methods of fining: physical and enzymatic.

A representative method of physical fining involves the use of Kakishibu. This fining method is based on the ability of persimmon tannin, the main component of Kakishibu, to form insoluble precipitates by binding with proteins. Kakishibu is used for fining sake because it demonstrates a stable fining effect across a wide pH range, without being affected by coexisting substances (Nakabayashi, 1968). If there are many turbid substances in the sake, adding Kakishibu alone can cause the aggregates to settle if the amount added is appropriate. However, generally, to facilitate the settling of aggregates, proteins such as gelatin and collagen, or polysaccharides derived from seaweed such as sodium alginate, are used in combination with Kakishibu. In the fining method using Kakishibu and proteins, the sake is first added with Kakishibu to form microscopic aggregates chemically bound with the turbid substances. However, these microscopic aggregates are usually too small to settle easily, so fining agents such as gelatin or collagen are further added to aggregate the microscopic aggregates, promoting their settling by forming larger aggregates. Thus, Kakishibu has excellent functionality as a fining agent for sake.

Our company first developed Kakishibu as a fining agent for soy sauce and seasoning liquids and began manufacturing and selling it in 1961. Soy sauce contains a large amount of turbid substances, so by simply adding Kakishibu, large aggregates are formed and settle, resulting in clear soy sauce. Subsequently, we developed and began manufacturing and selling Kakishibu "Mimasu Tamashibu" (product name) and gelatin as fining agents for alcoholic beverages.

Next, we developed a gelatin formulation to be used in conjunction with Kakishibu as a fining agent for alcoholic beverages. The development history is as follows.

In 1965, we received approval from the National Tax Agency under the product name "Oridaiya" and began manufacturing and selling it.

Subsequently, we developed a powdered Kakishibu formulation for alcoholic beverage fining agents, received approval from the National Tax Agency under the product name " Kakiraito," and began manufacturing and selling it. However, we later improved it to a granulated Kakishibu formulation. Then, in 1969, we began manufacturing and selling it under the product name "Granule Kakiraito."

In 1983, we established an improved method for producing liquid Kakishibu and began mass production of high-quality Kakishibu.

In 1984, the Kakishibu for alcoholic beverage fining agents underwent a product name change from "Mimasu Tamashibu" to "New Mimasu Tamashibu" and we began selling it.

In 2000, as an alcoholic beverage fining agent to be used in conjunction with Kakishibu, we newly developed a sodium alginate formulation and began manufacturing and selling it under the product name "Kiyosumi."

In 2001, we began manufacturing and selling the gelatin formulation "Oridaiya A" which uses fish collagen peptide as the main ingredient.

In 2002, we began manufacturing and selling the fish collagen formulation "Oridaiya M."

In 2005, we began manufacturing and selling the gelatin formulation "Oridaiya P [Granule]" which uses pig skin-derived collagen peptide as the main ingredient. This product, like "Granule Kakiraito," is granular, so it does not clump and dissolves easily when added to sake.

In 2008, we developed a vinegar fining agent "Kiyosumi S" to be used in conjunction with Kakishibu and began manufacturing and selling it.

Our Kakishibu fining agent is used in alcoholic beverages such as sake and mirin and in liquid foods such as soy sauce, vinegar, and honey.

The above-mentioned fining agents for alcoholic beverages and liquid foods have been continuously improved since the start of manufacturing and sales.

3.2.3 Applications in Health Foods and Processed Foods

Our company began manufacturing and selling Kakishibu as a food additive in 1970. The products are "Asutorin," a powdered Kakishibu in which dextrin is dissolved in Kakishibu raw liquid and then the resulting liquid is dried, and "Asutorin P," a powdered Kakishibu in which Kakishibu raw liquid is directly dried.

Meanwhile, in their experiments, Takeo Mimasu, one of the authors, and his colleague Kuniji Torii found using the powdered juice of astringent persimmons that Kakishibu has a strong deodorizing effect in 1984 and this achievement was patented in 1991 (Mimasu & Torii, 1991). In their experiments, each of the four types of standard malodorous gases recognized as such was taken into a 5-liter sealed container, and 1 g each of the active ingredient of this invention, persimmon fruit juice powder, was added. After 10 min, the gas in the container was taken out, the residual concentration was measured by gas chromatography, spectrophotometry, etc., and the removal rate was calculated by comparison with the control (Table 3.3).

From these experimental results, it became clear that "Asutorin" and "Asutorin P" are products with deodorizing effects, and our company was selected for the "Kyoto Small and Medium Enterprise Excellent Technology Award" in the fiscal

Table 3.3 Removal rate of various malodorous gases

Malodorous gas	Initial concentration (ppm)	10 min after addition concentration (ppm)	Removal rate (%)
Trimethylamine	2,250	0	100
Ammonia	2,500	0	100
Hydrogen sulfide	1,050	70	94
Acetaldehyde	75	12	84

year 1999 for the development of "Asutorin" deodorant formulation, which uses persimmon juice as an active ingredient. This technology award is given by Public Interest Incorporated Foundation Kyoto Industrial Support Organization 21 to small and medium-sized enterprises and technicians who have successfully developed excellent technologies or products befitting the traditional city of Kyoto. Our company, Mimasu Kashichi Shouten Co. Ltd., was one of the smallest companies among the award-winning companies, but Takeo Mimasu later said that he was very happy to hear from the executives at the award ceremony, "You have achieved great results in a small company."

Kakishibu has many functionalities, including aggregation action with proteins, antioxidant action, blood pressure-lowering action, blood sugar-lowering action, antibacterial and antiviral action, hangover improvement, and deodorizing action. Therefore, "Asutorin" and "Asutorin P" have been used for a long time as raw materials for health foods and processed foods since their release. Kakishibu products used as raw materials for alcoholic beverages and foods are mostly fining agents for alcoholic beverages such as sake and mirin, and liquid foods such as soy sauce, and there are not many Kakishibu products used as ordinary food additives other than "Asutorin" and "Asutorin P," which are highly valued by food manufacturers. For example, it is used to adjust the texture and hardness of udon, providing it with *koshi*, a firm texture that leads to a bite. It is also used in foods such as candies and gums for the purpose of mouth refreshment.

Since December 2019, the new coronavirus (SARS-CoV-2) infection (COVID-19) has become a global pandemic, and it was reported that Kakishibu could inactivate the new coronavirus (Furukawa et al., 2021). In a test tube comparison of "new coronavirus and saliva only" with "new coronavirus, saliva with added Kakishibu," it was found that the new coronavirus could be inactivated by adding Kakishibu. The authors believe that the inactivation of the new coronavirus by Kakishibu may also occur in the human oral cavity, but for inactivation, the concentration of Kakishibu and the contact time with the virus are extremely important to consider. Interest in foods such as candies and throat lozenges containing Kakishibu, and orders for our "Asutorin" and "Asutorin P," increased from food manufacturers.

For the broader maintenance of health, it is important to balance the intake of the five flavors in our diet: sourness, bitterness, sweetness, spiciness, and saltiness. Astringency is classified under sourness (Tatsumi, 2006). However, it is not easy to incorporate astringency into our daily diet. Therefore, with the aim of supplementing astringency, our company began manufacturing and selling a health food product made solely of Kakishibu called "Kaki no Sei" in 1973. Later, we also started manufacturing and selling "Kaki no Sei [Granule]" a granulated version of the liquid "Kaki no Sei." This product is a more palatable version of the "Kaki no Sei" with a robust astringency, sweetened for easier consumption. These Kakishibu health food products had a high demand after their release, but sales decreased as various other health food products appeared on the market during the health food boom. However, they continue to maintain stable sales, supported by consumers with a preference for natural products.

3.2.4 Use of Cosmetics

In 1967, the "Japanese Standards of Cosmetics Ingredients" were announced, establishing 44 test methods and 114 ingredients (Takahashi, 2007). This marked the beginning of Japan's official standards and test methods for cosmetic ingredients. Furthermore, *Keshohin Shubetsu Haigo Seibun Kikaku* (*Shohaiki,* in short), the Japanese cosmetic product ingredient specifications, was established in 1994, and revisions and additions of new ingredients continued until 1999. Persimmon tannin, the main component of Kakishibu, defined as "tannin obtained from Kakishibu liquid" in the *Shohaiki*, was listed as a powder.

In 2001, the cosmetic approval system was essentially abolished, and a system for displaying all cosmetic ingredients was introduced. This system allows ingredients other than prohibited and restricted ones to be formulated after safety confirmation under corporate responsibility. Accompanying this, prior standards were abolished in 2001, and the *Shohaiki* was abolished in 2006. With the relaxation of these cosmetic regulations, resulting in less regulatory oversight, it became easier for cosmetic manufacturers to use cosmetic ingredients under their own responsibility.

The "Asutorin P" as a food additive was similar in description and specifications to persimmon tannin in the *Shohaiki* and thus began to attract attention from cosmetic manufacturers as a cosmetic ingredient. Especially after the patent related to the deodorizing effect of Kakishibu (Mimasu & Torii, 1991) expired in 2004, the "Asutorin P" began to be widely used in cosmetics (e.g., soaps, mouthwashes, hair dyes, toothpaste), against the backdrop of increasing demand and requests for deodorant products. Soaps containing Kakishibu, in particular, became popular.

3.2.5 Safety and Assurance Cultivated Through Product Development for Alcoholic Beverages and Food

"Is it safe to put Kakishibu in sake and soy sauce?" Our company's second president, Kashichi Mimasu, was said to have expressed such concerns to his son, Takeo. Since the use of Kakishibu had been mostly for coating and dyeing, such concerns were quite natural. However, for the young Takeo, who believed that relying solely on coating and dyeing would limit the future of his business, the use of Kakishibu as a fining agent for sake and soy sauce was very appealing. Until then, there had been no consideration for hygiene in the production of Kakishibu, but he energetically improved the manufacturing environment and methods, enabling the production of high-quality Kakishibu suitable for use in alcoholic beverages and food. Improvements in hygiene management and manufacturing methods for Kakishibu have continued to this day.

At our company, we manufacture Kakishibu for coating and dyeing in essentially the same way as Kakishibu for alcoholic beverages and food, contributing to improved safety and assurance. The liquor fining agent, "New Mimasu Tamashibu," is used not only as a fining agent for alcoholic beverages by sake manufacturers but also for painting wooden parts such as ceilings, beams, pillars, walls, doors, and scaffolding inside sake breweries. This product is a certified liquor fining agent and also produces the unique brown color of Kakishibu, making it a safe and reliable paint.

3.2.6 Specifications and Standards for Kakishibu in Alcoholic Beverages, Food, and Cosmetics Applications

For fining agents used in alcoholic beverages such as sake and mirin, the National Tax Agency (1961, 1997) has established specifications and test methods, and the Conference on the Specifications of Sake Treating Materials has also established the "Specifications of Sake Treating Materials (1979)." As Kakishibu is listed in these specifications and test methods, Kakishibu used as an alcoholic beverage fining agent must meet these specifications.

Food additives are classified into several categories: (1) "designated additives," which are designated by the Minister of Health, Labour and Welfare after confirming their safety and effectiveness, (2) "existing food additives," which have been used for many years as natural additives, (3) "natural flavoring agents," and (4) "ordinary foods used as food additives." Kakishibu belong to the category of "existing food additives." Japan's Specifications and Standards for Food Additives (JSFA) define the specifications for the components of food additives, the standards for manufacturing, and the methods for ensuring quality and include both "designated additives" and "existing food additives." However, many existing food additives, including Kakishibu, are not yet listed in the JSFA. Therefore, the Japan Food Additives Association has created voluntary specifications for existing food additives not listed in the JSFA and has published them as "the Voluntary Specifications of Existing Food Additives," which can be purchased on the Japan Food Additives Association website https://www.jafaa.or.jp/publish. The addition of unlisted items to the official JSFA is progressing with each revision of the JSFA, and it is expected that Kakishibu will be listed in the JSFA in the near future. The specifications for our food additives, "Asutorin" and "Asutorin P," are based on the specifications for Kakishibu as an alcoholic beverage fining agent and the Voluntary Specifications of Existing Food Additives and have been accepted by users for a long time.

For cosmetic ingredients, the *Shohaiki*, in which persimmon tannin was listed, has been abolished, and as mentioned previous Kakishibu can now be used in cosmetics under corporate responsibility after confirming its safety. The ingredients that were listed in the *Shohaiki* were included in the Japanese Standards of Quasi-Drug Ingredients 2006, and this standard was revised in the Japanese Standards of Quasi-Drug Ingredients 2021, which is available on the webpage of the Pharmaceuticals and Medical Devices Agency https://www.pmda.go.jp/

review-services/drug-reviews/about-reviews/q-drugs/0004.html. Persimmon tannin, which had been standardized as a cosmetic ingredient, has changed its classification to a quasi-drug ingredient, but the content of the standards has not changed from the *Shohaiki*. It should be noted that quasi-drugs are defined as having a mild effect on the human body and are uniquely positioned between cosmetics and pharmaceuticals in Japan.

3.2.7 Various Kakishibu-Dyed Products Adapted to Modern Life

In an era dominated by plastic, there is a growing trend towards environmentally friendly and body-friendly natural materials. Therefore, we decided to develop Kakishibu-dyed products that are suitable for modern life, based on everyday items using Kakishibu that have supported people's lives for a long time before the plastic era, and incorporating new knowledge. The development and sale of Kakishibu-dyed products was mainly handled by our group company, "Kakishibu House Mimasu" (established by Takeo Mimasu in 2006). Since cotton dyed with Kakishibu is common in Kakishibu dyeing, we developed products such as sheets, mattress pads, pillow covers using Kakishibu-dyed cotton cloth and cotton socks. The features of Kakishibu-dyed cotton are:

1. *Coolness*:
   Kakishibu-dyed cotton fabric provides a refreshing feeling throughout the year, and especially a cool feeling in the summer, helpful during the humid summers in Japan. Focusing on knowledge and experiences passed down from our ancestors, we have created prototypes of Kakishibu-dyed sheeting and mattress pads, changing the material from Japanese paper to cotton cloth, which is easier to sew and procure. When we actually slept on them and conducted in-house tests, we were able to feel coolness and were once again moved by the natural power of Kakishibu. Next, we conducted an external survey on the usage of Kakishibu-dyed sheeting and mattress pads and received feedback from many people that the coolness helped them to sleep comfortably. Furthermore, in contact cooling sensation tests measured by the maximum heat absorption rate using a precision rapid thermal property measurement device, it was clearly shown that Kakishibu-dyed sheeting and mattress pads provide a cooling sensation.
2. *Odor prevention with anti-bacterial effects and deodorizing effect*:
   When adding odor prevention functions to fiber products, especially clothing, effectiveness against body odor is required. The leading causes of body odor are ammonia, acetic acid, and isovaleric acid. Sweat is odorless when it is in the sweat glands, but it produces a foul smell over time when it comes out onto the skin surface. This is because bacteria, such as Staphylococcus aureus, which are present on the skin, ingest and decompose sweat and sebum, and excrete odor-causing substances. Given the presence of a large number of Staphylococcus

aureus on the skin surface, materials with both antibacterial effects against *Staphylococcus aureus* and deodorizing effects against the resulting foul odors are particularly suitable for fiber products like clothing (Mizutani et al., 2013, 2017). Therefore, we examined the antibacterial properties of Kakishibu-dyed cotton cloth against Staphylococcus aureus (we used NBRC 12732 for the test), using the absorption method of "JIS L 1902: 2008." As a result, the number of live Staphylococcus aureus bacteria increased more than 600 times after 18 h of cultivation in undyed cotton cloth, while in Kakishibu-dyed cotton cloth, it decreased to less than approximately 1/800. From these results, it became clear that Kakishibu-dyed cotton cloth has a strong odor prevention with anti-bacterial effect.

We also examined the deodorizing effect of Kakishibu-dyed cotton cloth using the method defined by the Certification Standards of SEK Mark Textile Products (Japan Textile Evaluation Technology Council). We found that Kakishibu-dyed cotton cloth showed a strong deodorizing effect, with odor reduction rates of 99% or more for ammonia, 95% or more for acetic acid, and 97% for isovaleric acid, which were used as model foul odors.

These findings clearly illustrates that Kakishibu-dyed cotton cloth has both a strong odor prevention with anti-bacterial effect and a deodorizing effect. Due to these strong dual effects, washing of Kakishibu-dyed sheeting, mattress pads, and pillow covers are sufficient with normal water washing, and there is no need to wash frequently. Additionally, Kakishibu-dyed cotton socks are characterized by their refreshing and comfortable fit, and even if worn for 2–3 days in the humid and hot summer without washing, they do not smell. Typically, socks are not worn for consecutive days without washing. However, they can be worn the next day by simply washing them lightly with water and drying them, making them convenient for travel.

3. *Reinforcement and regeneration functions*:

   Kakishibu has been used to reinforce and rejuvenate fragile materials such as old pillars, boards, furniture, fabric, thread, and Japanese paper by applying or dyeing them. By repeatedly applying or dyeing over time, the materials can be kept durable indefinitely. A well-known example of fabric reinforcement and rejuvenation is the *sakabukuro*. This is a cotton bag used to press sake mash, dyed with Kakishibu to make it durable for repeated use. The *sakabukuro* in use were re-dyed each summer to rejuvenate them, thereby ensuring their durability and enabling long-term use. Kakishibu-dyed sheeting, mattress pads, pillow covers, and socks, like the *sakabukuro,* have reinforced fibers, which keeps them durable for a long time.

4. *Anti-virus effect*:

   The COVID-19 pandemic spread rapidly worldwide after December 2019. Wearing face masks became one of the effective measures to prevent infection. In this context, when it was announced in September 2020—before publishing a paper (Furukawa et al., 2021)—that Kakishibu has an inactivating effect on the new coronavirus, attention was focused on Kakishibu-dyed face masks, and there was a significant increase in demand (Fig. 3.4).

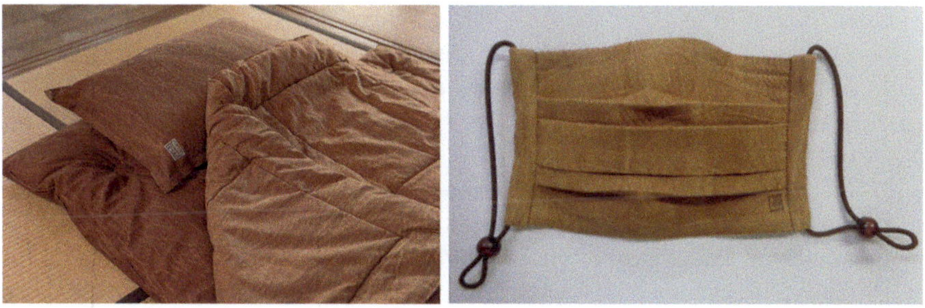

Fig. 3.4 Kakishibu-dyed beddings (left) and a Kakishibu-dyed face mask (right)

Kakishibu-dyed sheeting, mattress pads, pillow covers (Fig. 3.4), and socks were certified as "Kyoto Eco-Style Products" by the Kyoto Industrial Eco-energy Promotion Organization in the fiscal year 2012 because the following four points were evaluated. (1) The cooling sensation in summer reduces dependence on air conditioning, contributing to energy savings. (2) Kakishibu-dyed items are not easily prone to dust and have odor prevention with anti-bacterial and deodorizing effects. Therefore, washing usually only requires a simple rinse, reducing the frequency of detergent use. Furthermore, reducing the number of washes contributes to environmental protection and energy savings. (3) The product has a long lifespan and can be re-dyed and reused, which helps lower costs and reduce waste. (4) Products made from natural materials like cotton and dyed with Kakishibu can gently heal the body with nature's blessings. It is similar to cooling off on a riverside terrace with the breeze from the river during Kyoto's hot, humid summers. These products are connected to the Kyoto way of using nature wisely in daily life.

In addition to these, we have also commercialized Kakishibu-dyed comforters, mattresses, living mats, *zabuton,* Japanese cushion for sitting on tatami mats and the floor, *zabuton* covers, cushion covers, T-shirts, underwear, work clothes, aprons, stoles, scarves, face masks, sunshade arm covers, dishcloths, towels, bath mats, and bags.

Introduction of Other Companies' Initiatives

- The Niigata Prefecture Raw Noodle Industry Cooperative and the Niigata Agricultural Research Institute, Food Research Center, have developed ramen noodles kneaded with Kakishibu (Japanese patent no. 3671175). The noodles exhibit an advantageous texture characterized by firmness and hardness, and they resist dissolution during boiling, thereby rendering them exemplary and widely utilized noodles.
- In the beverage industry, Ishii Bussan Co., Ltd. in Nara Prefecture has commercialized persimmon leaf tea infused with Kakishibu.
- Shimamoto (2016) of Hiroshima University, in joint research with Altan Co., Ltd., discovered that Kakishibu had strong anti-norovirus activity.

(continued)

They succeeded in developing a disinfectant that is effective and safe against norovirus. This disinfectant, which has been on sale since 2007, is an excellent product that can also be used for hygiene management in cooking facilities.

- In 1990, Kanebo, Ltd. launched a makeup primer containing persimmon tannin and sold a significant amount. Persimmon tannin has strong skin-tightening and sebum secretion-inhibiting effects, improving the adhesion, durability, and finish of foundation. As a result, this cosmetic was recognized as a high-performance product (Kondo, 1992).
- Recently, many products such as shampoos, body soaps, and deodorant sprays containing Kakishibu have been sold.

3.2.8 Increasing Efficiencies in Transportation and Distribution

Kakishibu is a liquid and therefore heavy. This results in much effort and high energy costs during transportation and distribution. Furthermore, Kakishibu occasionally undergoes gelation during storage, whereby a fluid substance solidifies into a jelly-like state and becomes insoluble, rendering it unsuitable for long-term storage. The gelled substance must be discarded, which is a hindrance to use.

To solve these problems, we have improved the form of Kakishibu from a liquid to a powder, and further to a granular form, achieving lighter weight, smaller volume, more stable quality, and long-term storage. This has enabled significant reduction in energy costs during transportation and distribution, greatly improved ease of use, and eliminated unnecessary waste. Powder and granular products can significantly reduce transportation costs compared to liquid Kakishibu, making them advantageous for air transportation and increasingly used overseas.

The first developed powdered Kakishibu formulation "Kakiraito" was lightweight and reduced the burden of work, but it was unpopular because it clumped and was hard to dissolve. The further researched granular Kakishibu formulation "Granule Kakiraito" dissolves smoothly without clumping when put into the tank, producing a clear sake, and has become a popular product highly appreciated by sake manufacturers.

Furthermore, food additives usually require long-term storage, so liquid Kakishibu was difficult to use for health foods and processed foods. By turning the liquid Kakishibu into a powder, a more stable quality and long-term storage were achieved, making it easier to use for the same purposes. Also, unlike liquid Kakishibu, it can be mixed in its powdered form, and if dissolved in water, it can also be used in liquid form, expanding the range of usage methods.

Finally, "Mimasu Granule Kakishibu" (Fig. 3.5) is a product that has significantly reduced transportation costs by making liquid Kakishibu granular, lightweight, and small in volume, and because it is granular, it does not gel. It can easily

Fig. 3.5 Mimasu Granule Kakishibu

be returned to Kakishibu liquid by dissolving it in water, so you can use only the necessary amount when you need it, and the rest can be stored in granular form. This product was certified as a "Kyoto Eco-Style Products" (Kyoto Industrial Eco-energy Promotion Organization) in the fiscal year 2008, one of the first certified products in this system.

3.3 Examples of Functional Enhancement by Combining Kakishibu with Natural Substances

Along with its many individual functions, Kakishibu has also been used in combination with various organic and inorganic natural substances to enhance its functions and broaden its range. Some examples include:

3.3.1 Hull Paint as the Japan's First Patent: Combination of Kakishibu and Urushi

The following R&D story is mainly from an article from the website of Nippon Kako Toryo Co., Ltd. founded by Zuisho Hotta https://www.nippon-kako.co.jp/patent.html and Onuma (2010).

On August 14, 1885, Zuisho Hotta obtained Japan's first patent for an anti-rust paint and its application method. Kakishibu was used as one of the paint components. At that time, Hotta, known as a sculptor and lacquer artist, was inspired by a conversation among government officials. The conversation was about how the hulls of iron ships around the world were being corroded by seawater, requiring them to be dry-docked and repainted every six months. If a more potent anti-rust paint could be developed to extend the dry-docking cycle, then it would be of great benefit not only to our country but also to the world. The hull of an iron ship, submerged below the load waterline, faces harsher conditions than typical onshore coated materials. Salt causes severe rust corrosion, and marine organism fouling reduces vessel speed and wastes fuel. Antifouling was a challenge since the era of wooden ships, but with the advent of iron ships, the development of anti-rust function became a new important issue. Hotta, knowing the anti-rust function of *urushi*, Japanese lacquer that had been applied to armor and Nanbu iron kettles since ancient times, began researching hull paint with *urushi* as the main ingredient. After repeating undersea immersion experiments with various painted iron pieces and conducting real ship tests, he succeeded in developing a hull paint with superior anti-rust function compared to Western paints. Although the title of the patent is anti-rust paint, it actually had both anti-rust and anti-fouling functions. The patented paint comes in four types, all of which use *urushi* as the main ingredient, to which other materials such as iron powder, red lead, and oil soot are added, along with Kakishibu, ginger, and vinegar. By using red lead as an anti-rust pigment and generating iron tannin through the reaction of persimmon tannin, he aimed to strengthen the paint film and improve its anti-rust power. It is also believed that ginger was used in combination to provide anti-algae and anti-shellfish properties due to its phenolic compounds. Iron powder is mixed in a weight second only to *urushi*, aiming to relieve the stress of the hardened film, prevent film cracking, and improve interlayer adhesion. The painting method involves applying these four types of paint in layers and spraying steam when the humidity is low and drying is slow. By layering the four types of paint to form a multi-layer structure, the stress relief of the entire hardened paint film and the prevention of film cracking are achieved.

In 1887, Hotta's anti-rust paint was officially adopted by the Navy. It was painted not only on warships but also on civilian ships and even on Russian warships. Compared to general oil-based paints that need to be repainted every six months, Hotta's anti-rust paint does not rust for three years, so despite the high docking fees due to heating, humidifying, and long drying times, it was calculated to be cheaper in total cost and received high praise from overseas.

In 2009, an experiment to reproduce the effects of Hotta's anti-rust paint was conducted. As a result of a six-month undersea immersion test of the painted board, it was found to be inferior to modern hull paints in terms of the adhesion of marine organisms such as barnacles, but it successfully protected the iron plate from corrosion. Also, from this immersion test, a very interesting antifouling mechanism was revealed, where marine organisms adhere and accumulate to a certain thickness, the upper layer of the layered paint film peels off with the marine organisms, and the lower layer of the paint film is exposed as a new antifouling surface.

Hotta's anti-rust paint can be said to have been the cutting-edge technology of the time, excelling in the anti-rust function of iron parts and reducing the adhesion of barnacles and seaweeds to the hull. In the era around 1887, when Japan's paint industry was still immature, Zuisho, with his progressive ideas and enthusiasm, developed a hull paint that attracted worldwide attention using only domestic raw materials, including *urushi*. One cannot help but feel the enthusiasm of the researchers of the Meiji era, who were striving for modernization.

3.3.2 Architectural Paint: Combination of Kakishibu and Oil Fumes and Red Iron Oxide

As mentioned in Sect. 3.1.5, Kakishibu was used as an architectural paint. Kakishibu was not only used alone, but it was also often used in conjunction with pigments such as oil fumes and red iron oxide to combine the water-resistant and preservative effects of Kakishibu with the coloring effect and preservative effects brought by oil fumes and red iron oxide. Currently, Kakishibu also works to fix oil fumes and red iron oxide, making it difficult to remove from the painted surface.

3.3.3 Cultivation Using Iron with Tannin: Combination of Kakishibu and Iron

Iron in nature quickly reacts with oxygen to form iron oxide, which plants hardly absorb as is. However, when iron is combined with tannin, it becomes easily absorbed and utilized by plants. Iron with tannin can be readily made from materials containing tannin, such as green tea, Kakishibu, chestnut tannin skin, sawtooth oak leaves, and iron materials like iron nails and cast iron. When spread in the fields, iron with tannin is known to have various effects, including promoting larger, tastier crops that are more resistant to disease (Rural Culture Association Japan, 2023; Nonaka, 2021).

3.3.4 Adhesive: Combination of Kakishibu and Glue

When pasting several sheets of *washi*, Japanese paper, together, applying Kakishibu can strongly adhere each sheet of *washi*. When making a Japanese umbrella, Kakishibu is mixed with glue (a mixture of half tapioca and bracken powder) to firmly attach the Japanese paper to the bamboo bones. (Imai, 2003)

3.3.5 *Brewing Fining Agent: Combination of Kakishibu and Gelatin, Collagen, and Polysaccharides*

As already described in Sect. 3.2.2, the combination of Kakishibu and proteins such as gelatin, collagen, and polysaccharides derived from seaweed such as sodium alginate can make fining of sake easier.

3.4 Future Prospects

Chemically synthesized products, including plastics, are designed and optimized for their chemical structures to improve their desired functionality. On the other hand, the history of Kakishibu began during a time when resources were scarce, and it was tested for various practical purposes in daily life because it was readily available. Its usefulness and multifunctionality were gradually discovered. However, since Kakishibu was not made for any purpose, its functionality is inferior compared to chemically synthesized products. Therefore, Kakishibu was replaced by chemically synthesized products such as plastics. However, non-natural substances such as plastics often have low biodegradability, causing environmental problems such as microplastics. On the other hand, natural substances including Kakishibu are biodegradable and incorporated into the natural cycle, making them safe and reliable materials. Therefore, in the future, it will be important to develop new natural materials that replace chemically synthesized products such as plastics by, for example, combining Kakishibu and natural substances as mixtures, maintaining biodegradability, and improving their functions or expanding the range of functions as seen in Sect. 3.3.

While writing the history of Kakishibu, we felt the passionate thoughts and tireless efforts of our predecessors in making Kakishibu and using it to create things. Our company aims to develop products suitable for modern life by using new science, technology, and information to build on the valuable experiences of those who came before us.

Additionally, we want to share the usefulness and appeal of Kakishibu with people abroad and to showcase the charm of Kakishibu to an international audience.

References

Aikenhead, G. S., & Ogawa, M. (2007). Indigenous knowledge and science revisited. *Cultural Studies of Science Education*, 539–620. https://doi.org/10.1007/s11422-007-9067-8

Akiyama, H., Uchiyama, K., Koide, I., Himeno, K., & Noda, Y. (1969). Seishu seichoyo kakishibu no hinshitsu ni tsuite (Dai 1 ho): Kakishibu no seibun ni tsuite (Quality of kakishibu for purifying sake (First Report): Components of Kakishibu). *Journal of the Brewing Society of Japan, 64*(10), 889–893.

Arakawa, M. (1964). Joso gono seishu ni tsuite (About sake after pressing). *Journal of the Brewing Society of Japan, 59(4)*, 283–286.

Brewing Society of Japan. (1993). *Dai 4 kai Kaisei Kokuzeicho Shotei Bunsekiho Chukai (Commentary on the fourth revised National Tax Agency designated analysis method)*. Brewing Society of Japan.

Conference on the Specifications of Sake Treating Materials. (1998). *Nihonshuyoshizai Q&A (Sake treating materials Q&A)*. Brewing Society of Japan.

Ehime Kenshi Hensan Iinkai Shakai Keizai Ichi Bukai (Ehime Prefectural History Compilation Committee, Social and Economic Affairs Subcommittee I). (1985). *Showa koki no senkai yoshoku (Shallow sea aquaculture in the late Showa period)*. Ehime Kenshi Shakai Keizai 2 Nourinsuisan (Ehime Prefecture History, Social Economy 2, Agriculture, Forestry and Fisheries).

Fujiki, Y. (1940). Koden no nori to bocyuzai ni tsuite (About traditional glue and insect repellent). *Shomotsutenbou (Book Outlook), 10*(5), 408–410.

Fujita, Y. (2002). I-4-(2) Gyogyo: Nori yoshoku (Fisheries: Nori (edible laver) cultivation). *Nippon Suisan Gakkaishi, 68*(1), 102–103.

Furukawa, R., Kitabatake, M., Ouji-Sageshima, N., Suzuki, Y., Nakano, A., Matsumura, Y., Nakano, R., Kasahara, K., Kubo, K., Kayano, S., Yano, H., & Ito, T. (2021). Persimmon-derived tannin has antiviral effects and reduces the severity of infection and transmission of SARS-CoV-2 in a Syrian hamster model. *Scientific Reports, 11*, 23695. https://doi.org/10.1038/s41598-021-03149-3

Hyojunbunsekiho chukai Iinkai (Commentary on Standard Analytical Methods Committee). (2017). *Shuruisogokenkyusho hyojunbunsekiho chukai (Commentary on Standard Analytical Methods of National Research Institute of Brewing, Japan)*. Brewing Society of Japan.

Imai, K. (2003). *Kakishibu: Mono to Ningen no Bunkashi (Kakishibu: Cultural history of things and humans)*. Hosei University Press

Inoue, K. (2007). Development of the technology for removing toxic metals and recovering valuable metals by the effective use of biomass wastes. *Journal of MMIJ, 123(2)*, 59–67.

Inoue, K. (2014). Development of adsorptive removal technology of radioactive elements from environment sing persimmon tannin and astringent persimmon wastes. *New Food Industry, 56*(11), 8–16.

Itamura, H., Sun, N., Niimura, M., Shimosaki, S., Nakatsuka, A., Ando, H., Nagamatsu, H., Katsube, T., Tsurunaga, Y., Asano, Y., Arakawa, O., Kamei, T., Murata, K., Suginaka, K., Funaki, K., Matsumoto, T., Kojima, Y., Uemura, E., Murakami, N., & Matsuzaki, H. (2011). Effect of ingestion of the Japanese persimmon 'Saijo' fruit on ethanol levels in the blood of humans and rats. *Food Preservation Science, 37*(4), 155–165.

Japan Food Additives Association. (2021). *Dai 5 han Kisontenkabutsu Jishukikaku* (5th ed.). The Voluntary Specifications of Existing Food Additives.

Kanzaki, S. (2016). The origin and cultivar development of Japanese persimmon (*Diospyros kaki* Thunb.). *Nippon Shokuhin Kagaku Kogaku Kaishi, 63*(7), 328–330.

Kawamura, Y. (2006). *Ariakekai Saiseikiko, Ariakekai Koza Koenroku* (Ariake Sea Restoration Organization Ariake Sea Lecture Record).

Kida, S. (1963). Joso, oribiki, seicho no jittai ni tsuite (The actual process of pressing, sediment removal, and fining). *Journal of the Brewing Society of Japan, 58*(3), 200–202.

Kikuchi, A. (1948). Kajushuruikakuron, dai 14 sho kaki (Fruit tree species, chapter 14 persimmon). In *Kajuengeigaku Jokan (Fruit Tree Horticulture volume 1)* (pp. 347–400). Yoken-do.

Kimishima, S., & Fujii, S. (1966). Fukuroko no jokyo ni tsuite (Removing bag odor). *Journal of the Brewing Society of Japan, 61*(6), 526.

Kimura, T. (1966). Hiire to sono zengo no chui (Pasteurization and precautions before and after pasteurization). *Journal of the Brewing Society of Japan, 61*(4), 303–306.

Kofujita, H. (2013). *Studies on biodegradation of condensed tannin*. Kakenhi Basic Research (C) report.

Kometani, T., & Takemori, K. (2016). Polyphenols from persimmon fruits as a functional foods Material. *Nippon Shokuhin Kagaku Kogaku Kaishi, 63*(7), 331–337.

Kondo, M. (1992). Saikin no awajo sukin kea seihin no kenkyu kaihatsu/Kanebo kabushikigaisha keshohin kenkyusho (Recent research and development of foaming-cosmetics for skin care/ Cosmetics Laboratory, Kabebo, Ltd.). *Fragrance Journal*, 55–60.

Kuruma, M. (1975). Seisei hiire (Purification process, pasteurization). *Journal of the Brewing Society of Japan, 70*(8), 541–545.

McCord, J. M., & Fridovich, I. (1969). Superoxide dismutase: An enzymic function for erythrocuprein (hemocuprein). *Journal of Biological Chemistry, 244*(22), 6049–6055. https://pubmed.ncbi.nlm.nih.gov/5389100/

Mimasu, T., & Torii, K. (1991). *Kaki no sakujueki o yukoseibun to shita shoshu seizai (Deodorant formulation with persimmon juice as an active ingredient)* (Japan Patent No. Hei 3–61457). (Published September 19, 1991; registered as JP Patent No. 1707185 on October 27, 1992.)

Mena, E., & Nunokawa, Y. (1971). Seishu seichoyo kakishibu ni kansuru kenkyu (Dai 1 ho): Koroido tekitei ni yoru kakishibu no teiryo ni tsuite (Research on kakishibu for fining sake (First Report): Quantification of kakishibu by colloid titration). *Journal of the Brewing Society of Japan, 66*(6), 620–623.

Miyagi Ken Gyogyo Kyodokumiai (Miyagi Prefecture Fisheries Cooperative Association). (2025). *Nori Yoshoku (Nori Cultivation)*. Retrieved February 16, 2025, from https://www.jf-miyagi.com/omotenashi/html/zukan14b.html

Mizutani, C., Yahata, A., Tsuiki, H., Morikawa, H., Kajiwara, K., Takahashi, K., Shigeta, T., Kurosawa, H., Otsuka, C., & Shirai, H. (2013). Application of nursing care using deodorant and antibacterial fibers. *Sen'i Gakkaishi, 69*(7), 141–145.

Mizutani, C., Kawanoue, Y., Hirano, Y., & Hirota, R. (2017). Odor of sportswear made of polyester and its deodorant/antibacterial processing: Effect of acid-processed polyester fabric on skin pH and odor. *Desanto Supotsu Kagaku (Descente Sports Science), 38*, 10–16.

Nakabayashi, T. (1968). Kakishibu no kagaku (Chemistry of kakishibu). *Journal of the Brewing Society of Japan, 63*(11), 1149–1154.

Nakabayashi, T. (1971). Kakishibu no dekirumade (How Kakishibu is made). *Journal of the Brewing Society of Japan, 66*(5), 474–476.

National Research Institute of Brewing. (2018). *Osakenohanashi, tokushu: Wain 1 (gaiyo) (Talk about sake, special feature: Wine 1 (overview))*. National Research Institute of Brewing Information Magazine, 2.

National Tax Agency. (1961). *Kokuzeicho Shotei Bunsekiho* (National Tax Agency Designated Analysis Method).

National Tax Agency. (1997). *Shurui hozon no tame shurui ni konwa surukotogadekiru buppin no toriatsukai ni tsuite (Horei kaishaku tsutatsu)* (Handling of Items that can be mixed with alcoholic beverages for preservation). Frequently revised and the latest version was published in 2024.

Nonaka, T. (Inventor). (2021). *Tetsubun kyokyu zai oyobi seibutsu ikusei hoho* (Iron supply agent and biological cultivation method) (Japan Patent No. 6908940). Ippan Shadan Hojin Tetsu Mineraru (General Incorporated Association Iron Mineral).

Ogata, S. (1976). Kakinomi no arukoru meitei eno koka (The effect of persimmon fruit on alcohol intoxication). *Journal of the Brewing Society of Japan, 71*(7), 488–495.

Ogata, S. (1986). *Kaki to kenko: Kakishibu no koyo (Persimmon and health: The benefits of Kakishibu)*. Mimasu Kashichi Shouten Co., Ltd.

Oguri, T. (2008). Development of soy sauce manufacturing technologies. *Gijutsu no Keitoka Chosa Hokoku (Systematization of Technology Survey Report), 10*, 131–208.

Ohgai, M. (1986). Studies on the morphology and ecology of epiphytic and epilithic diatoms in nori cultivation grounds. *The Journal of the Shimonoseki National University of Fisheries, 34*(2·3), 170–222.

Onuma, K. (2010). Systematic survey of the development of paint technologies. *Gijutsu no Keitoka Chosa Hokoku (Systematization of Technology Research Report), 15*, 3–78.

Oto, K., Inoue, K., Hirota, K., & Parajuli, D. (2012). Baiomasu haikibutsu wo yukoriyo shita shiyozumi kogata kaden seihin karano reametaru kaishu gijutsu no kaihatsu (Development of rare metal recovery technology from used small household appliances utilizing biomass waste) (K2131, K22075, K2354). *Heisei 23 nendo (Fiscal Year 23) Kankyo Kenkyu Sogo Suishinhi Hojokin Kenkyu Jigyo Sogo Kenkyu Hokokusho*. Environmental Research Comprehensive Promotion Fund Subsidy Research Project Comprehensive Research Report.

Rural Culture Association Japan. (2023). *Tetsu tokoton katsuyo dokuhon* (Iron utilization handbook). Ippan Shadan Hojin Nosangyoson Bunka Kyokai (General Incorporated Association Rural Mountain Fishing Village Cultural Association).

Sakaguchi, T. (1996). *Bioaccumulation of Uranium*. Kyushu University Press.

Shimamoto, T. (2016). Nipponbunka ni nezuita kakishibu no kagaku (The chemistry of kakishibu, deeply rooted in Japanese culture). *Chemistry & Education, 64*(7), 348–349.

Suda, I. (2000). Bunkougakuteki kosanka kino hyoka (Spectroscopic evaluation of antioxidant function). In K. Shinohara, T. Suzuki, & S. Kaminogawa (Eds.), *Shokuhin Kino Kenkyu Ho (Food function research method)* (pp. 218–220). Korin.

Tajima, K. (2016). A systemized exploration of petrochemical technology. *Gijutsu no Keitoka Chosa Hokoku (Systematization of Technology Survey Report), 23*, 117–256.

Takahashi, M. (2007). Mini fairu: Genzairyo no kikaku to bunseki ho: iyaku bugaihin keshohin (Mini file: Raw material specifications and analysis methods: Quasi-drugs & cosmetics). *Bunseki (Analysis), 29–30.*

Tanaka, M. (1991). Shokubutsu wo mochiita edojidai no shoseki gaichu bojoho (Plant-based methods of pest control in books during the Edo period). *Kaokugaichu (Household Pests), 13*(2), 86–90.

Tanaka, T. (1997). Kobunshi shokubutsu porifuenoru (Polymeric plant polyphenol). *Polymers, 46*(12), 893.

Tatsumi, N. (Ed.). (2006). *Yakuzen sozai jiten (Dictionary of medicinal ingredients)*. Gensosha Ltd..

Tochikura, T. (1994). *Zoho shoyu no kagaku to gijutsu (Science and technology of soy sauce)* (Expand ed.). Brewing Society of Japan.

Tsukita, T. (1984). *Kateiniokeru jissaitekikango no hiketsu: zohoshinteiban (Secrets of practical nursing at home)* (Expand New ed.). Kensukobunkan.

Ueda, K., Kawabata, R., Irie, T., Nakai, Y., Tohya, Y., & Sakaguchi, T. (2013). Inactivation of pathogenic viruses by plant-derived tannins: Strong effects of extracts from persimmon (*Diospyros kaki*) on a broad range of viruses. *PLoS One, 8*(1). https://doi.org/10.1371/journal.pone.0055343

Tomokazu Goto was born in Hyogo Prefecture. After graduating from the Department of Synthetic Chemistry at Nagoya Institute of Technology in 1981, he completed his master's course of studies at the Department of Synthetic Chemistry, Graduate School of Engineering, Nagoya Institute of Technology in 1983. He subsequently held the position of Senior Researcher and Team Leader at The Green Cross Corporation and Mitsubishi Pharma Corporation, where he was involved in drug discovery research for synthetic pharmaceuticals, primarily within the fields of inflammation, allergy, and immunology. From 1994 to 1996, he pursued further academic development abroad at the University of California, Irvine, under the supervision of Professor Larry E. Overman. In 2002, he joined Mimasu Kashichi Shouten Co., Ltd., and has served as the Head of the Research and Development Department.

Yoshitsugu Mimasu was born in Kyoto Prefecture. After completing his university education, he became a member of Mimasu Kashichi Shouten Co., Ltd. Subsequently, he served as Senior Managing Director before being appointed President in 2003. He has held this position continuously to the present day.

Takeo Mimasu was born in Kyoto Prefecture. After graduating from high school, he joined Mimasu Kashichi Shouten, which had not yet been incorporated as a corporation. In 1968, he founded Mimasu Kashichi Shouten Co., Ltd., and assumed the position of President. Subsequently, he has served as Chairman since 2003. Furthermore, in 2006, he established Kakishibu House Mimasu and has served as President.

Chapter 4
Science of Kakishibu: Traditional Knowledge and Western Science and Technology

Tomoaki Matsuo

Abstract This chapter aims to explain the science of Kakishibu in a way that is easy to understand even for those who have never encountered the word Kakishibu before. Beginning with an explanation of polyphenols related to the taste of green tea, it first describes the difference between these polyphenols and the astringent component of persimmons, condensed tannin, and then traditional uses of Kakishibu made from astringent persimmons which included some of tools used in traditional tea production and others, its physical and chemical properties of traditional Kakishibu, the various measurement methods of the characteristics, and how it differs from plastics made from coal and oil. It also covers its component analysis and the structure of its main polyphenolic component, kakitannin, and the entity is a huge size of proanthocyanidin polymer. It is believed that their knowledge and techniques will encourage many people to consider the new industrial possibilities and attractions of this natural high molecular substance, kakitannin. Perhaps it will provide us with clues to saving our beloved planet. The following sections are consisted in this chapter: a detailed explanation of the science behind Kakishibu. After a brief introduction to the science of Kakishibu (Sect. 4.1), the scientific explanation will progress from the macro to the micro level. It will delve from the substance of Kakishibu and products utilizing Kakishibu (Sect. 4.2) to the components contained within Kakishibu (Sect. 4.3) and finally to the molecular structure of Kakishibu's main component, kakitannin (Sect. 4.4).

Keywords Antivirus · Astringent persimmons · Catechins · Condensed tannin · Heavy metal binding · Kakitannin · Proanthocyanidin polymer · Protein precipitation · Tea polyphenol

---

T. Matsuo (✉)
The Scientific Research Association for Kakishibu and Kakitannin, Kyoto, Japan
e-mail: kaki@tannin.info

K. Kano et al. (eds.), *Traditional Knowledge for Green Innovation with Japanese Persimmon Tannin, Kakishibu*, https://doi.org/10.1007/978-981-95-7805-4_4

4.1 Introduction to Science of Kakishibu

4.1.1 Sparrow's Leaf Hideout: Kakishibu and Green Tea

In the medieval period of Japan, the Edo period, green tea cultivation and production were thriving in the southern region of Kyoto, and various technologies were rapidly developed. As a result, a sustainable natural material circulation system was established around tea cultivation. The symbol of these technologies was the persimmon tree. In the usual tea fields of that period, persimmon trees were scattered around and within the fields. While tea trees are evergreen, persimmon trees are deciduous. Why was it necessary to plant these two types of trees together at that time? (Fig. 4.1).

The astringent persimmon trees planted in tea fields generally seem to have played a role in windbreak, frost protection, and pest control, but what is more interesting is that they also served as indicators for farming work. Since persimmon trees are deciduous, they sprout buds in early spring, and tea farmers would observe the movement (growth status) of these buds, leaves, and twigs in order to know when to start various annual tasks. There is a traditional teaching that when a sparrow perches on a persimmon tree and is seen hiding in the shadow of the branches and leaves, spread *yoshizu*, the reed screen covering tea trees, and thereafter when a crow perches on the branches but is seen hiding, spread straws on the screen to make the sunshade more. If *tare*, the hanging sunscreen made of straw to cover the sides is bad, it is because of the way it is made. Observing the growth of the persimmon tree's leaves, farmers would start the task of making a roof of straw over the tea tree leaves to control the amount of sunlight exposure (Horii, 2006; Imai, 2021).

After this task, as the season progresses further, it becomes the time for tea picking. The quantity of light hitting the tea plant directly impacts the levels of polyphenols and umami constituents synthesized within the leaves, thereby significantly

Fig. 4.1 Astringent persimmon trees planted in a tea plantation in Kyoto, playing various roles—scenes of early summer and late autumn trees. These photos were provided by the Kyoto Prefectural Yamashiro Local History Museum, Kyoto

affecting the flavor profile of the final tea product. Also, the fruits of the astringent persimmon trees in and around the tea fields are harvested as green immature persimmon fruits at the height of summer and squeezed to make Kakishibu.

Furthermore, when they change color in the fall, some of the fruits are peeled after harvest, dried, and used as dried persimmons for food. The squeezed juice of the green immature fruits mentioned earlier undergoes fermentation and aging to become Kakishibu, which is used in the winter and early spring to make various tools used in tea production (*Nagapote*, Shibuutiwa, *Mi*, *Marupote*, Fig. 4.2), to repair damaged parts of them during use, and to make tea storage containers. Kakishibu was valued for its ability to form a highly water-resistant and durable film when applied to *washi,* a traditional Japanese hand-made paper selected as Intangible Cultural Heritage by UNESCO, and dried. These tools were consistently used so repairs were essential, and Kakishibu was needed every year. Kakishibu was necessary for various tools used in tea production (Fig. 4.2) and had several valuable properties, including excellent durability and resistance to humidity, improved operability due to good slip of tea leaves on the tea tools, and non-transferral of tea leaf scent (Kanbayashi & Kanbayashi, 2017; Kyoto Prefectural Yamashiro Local History Museum, 1997, 2009).

(a) (b) (c)

Fig. 4.2 Tools used in traditional green tea production. Those tools were made from bamboo framework, *washi*, and Kakishibu. (**a**): Nagapote: used to spread tea leaves in tea sorting. Shibuchiwa: used to create wind to cool and evaporate moisture from tea leaves. (**b**): Mi: used to spread tea leaves, with selection by weight. (**c**): Marupote: used as a container for tea leaves. (**a, b** photos were provided by Tuyoshi Matsushima of Kanbayashi Syunsho Honten Co., in Kyoto and C photo by the Kyoto Prefectural Yamashiro Local History Museum, Kyoto)

After squeezing the astringent persimmon fruit, the squeezed residue of persimmons (the pulp, skin, and stems left after squeezing the juice from persimmons) was naturally used as compost in the fields. A sustainable resource cycle was thus created around this unique natural material. For tea farmers at the time, the production of Kakishibu and dried persimmons also had the significance of increasing income through supplementary jobs.

When Japanese people bite into a persimmon, they describe it as astringent. When drinking tea, they also often savor it while saying it is astringent. Both are known to contain a considerable amount of polyphenols. However, the astringency of the two does not lead to the same sensation or taste. We will now explore the similarities and differences in astringent components, polyphenols, in persimmon and tea. Green tea and persimmon trees have the same order, Ericales, even though they have different families. What similarities and differences are there in the chemical components related to their astringency?

4.1.2 The Scientific Astringent Component of Persimmons

In order to defend themselves against excessive ultraviolet rays from the sun, microbe invasion, and damage from insects and small animals, plants synthesize and accumulate various defensive substances. Sometimes they suddenly produce these in response to special sensors (induced chemical defense), or they take time to produce them in large quantities for storage (constant chemical defense). Among these compounds, polyphenols are believed to be one of the substances produced for protection from various environmental stresses during the evolutionary process of plants spreading from the sea to the land. Currently, polyphenols are contained in most plants to a greater or lesser extent, and it is estimated that there are about 8000 types (Appel, 1993; Barbehenn & Constabel, 2011; Harborne, 1977; Konno, 1996; Scalbert & Haslam, 1987; Yamane et al., 2010).

The polyphenols initially produced by plants are simple in structure, consisting of only a few phenolic hydroxyl groups attached to a single benzene ring. For example, in addition to resorcinol, gallic acid found in oak bark and galls, caffeic acid present in coffee and resveratrol found in red wine and peanut skins are typical examples (Fig. 4.4). Initially these simple defensive substances appeared to have worked effectively as biological barriers, but surrounding invaders developed functions and abilities to breach these barriers over the course of evolution. In response, and through a process of co-evolution (Spencer, 1988; Zunjarrao et al., 2020), plants produced more complex polyphenols and developed more effective chemical defense mechanisms. The structure of polyphenols has become increasingly more complex, and complexification (esterification, condensation etc.) and polymerization occurred sequentially in their molecules (Fig. 4.4).

Currently, the leaves of the tea tree, an evergreen tree of the Rhododendron order, mainly synthesize and accumulate four types of polyphenols called catechins (also scientifically known as flavan-3-ols). Catechins, which have chemically evolved from simple polyphenols, have a special structure consisting of an A ring and a B ring with hydroxyl groups. From this basic structure, further complexity (e.g., addition of gallate ester) emerged, giving birth to four types of catechins (Fig. 4.4). These are epicatechin, epicatechin gallate, epigallocatechin, and epigallocatechin gallate. In the average tea leaf, the first tea of the year (the first pick of the new season) contains fewer of these four types of catechins and has milder astringency, but the second tea (the leaves picked after the initial harvest) shows increases in these amounts and the content ratio of them: (−)-epicatechin 1.5%, (−)-epicatechin gallate 4.1%, (−)-epigallocatechin 3.7%, (−)-epigallocatechin gallate 12.2% (Ishigaki, 1981; Saito, 2018). Although the analytical values differ significantly depending on the type of raw tea leaves and sencha green tea, some common points can be observed. The leaves contain a higher concentration of epigallocatechin compared to epicatechin, which possesses the simplest structure, as well as an increased presence of their gallate ester derivatives. This epicatechin has four phenolic hydroxyl groups on two benzene rings, while the most complex of the four, epigallocatechin gallate, has three benzene rings and eight hydroxyl groups (Fig. 4.4). It is known that this polyphenol, which has many adjacent hydroxyl groups, binds relatively strongly to proteins and exhibits astringency. Notably, the tannin of tea is a different polyphenol from the tannin of astringent persimmons. It had been speculated that the tannin of astringent persimmons is a type of polyphenolic polymers for a long time, and it has been proved. This will be explained in section 4.4.

Phenol

Phenol is a compound that has one phenolic hydroxyl group on a benzene ring. The first plastic made in the world was phenolic resin, but its raw materials were phenol and formaldehyde. This phenol is extracted from coal tar acquired through the dry distillation of coal and is produced in substantial quantities. In other words, phenolic resin is a completely synthetic chemical (Fig. 4.3).

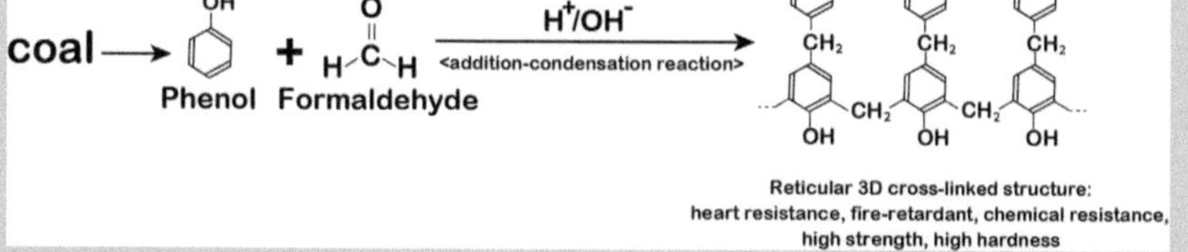

Fig. 4.3 Chemical structure of the first chemically synthesized phenolic resin (plastic) from coal-derived phenol

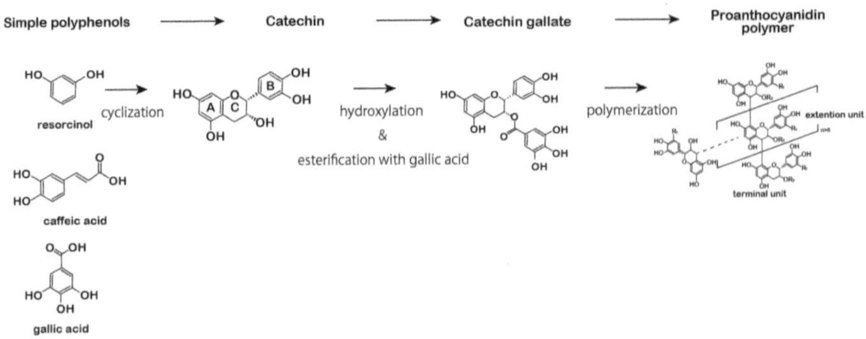

Fig. 4.4 Promotion of chemical structural evolution of polyphenols by co-evolution in plants, complexification and polymerization

Polyphenol

Polyphenol is a general term for compounds that have two or more phenolic hydroxyl groups on the benzene ring. It is a type of secondary metabolite produced by plants, and some of the polyphenols included in food ingredients may have astringent or bitter tastes. There are around 8000 types in the world. The term polyphenol is derived from the Greek word "poly," meaning many or numerous, and in this context, it refers to compounds that have many phenolic hydroxyl groups within the molecule.

Tannin

Tannin is a group within polyphenols and is a general term for high molecular weight (HMW) compounds that have an astringent taste. Common characteristics include an ability to strongly binding with many proteins and forming a blue-black color when combined with iron ions. The name tannin comes from "tan," which refers to the process of tanning raw hides with the hot water extract obtained from bark and heartwood chips. Since 1966, when Professor Haslam of Sheffield University classified tannins into hydrolyzable tannins and condensed tannins (non-hydrolyzable tannins), the name is not chemically accurate, but it has been used for a long time in practical situations. Hydrolyzable tannins are a general term for tannins that produce polyalcohols and polyphenolic carboxylic acids by hydrolysis with acid, alkali, or enzymes (tannase) (Haslam, 1979). Condensed tannins were once thought to be HMW compounds with unknown structures that were not hydrolyzed, but it is now clear that most of them are HMW compounds in which many structural units of flavan-3-ols (catechins) are repeatedly bonded, known as proanthocyanidin polymers (or polymeric proanthocyanidins).

Tannin and Astringency

The astringency of persimmons is completely different from the astringency of green tea and black tea and is similar to the astringency of red wine but more intense. Practically, they are all expressed as containing tannin, but chemically, their structures are very different. The astringency you sense when you bite into an astringent persimmon is not a taste but is caused by the formation of a complex between the tannin and human oral proteins. Normally, the surface of the oral cavity is protected by a slimy layer formed by proteins called mucin. The formation of this complex removes this slimy layer, resulting in a loss of smoothness on the surface, creating a rough or prickly, dry sensation (Bajec & Pickering, 2008; Murry et al., 1994; Wang et al., 2024).

The astringent components of persimmons accumulate specifically in special cells called tannin cells in the fruit. This can be observed by utilizing various color reactions of tannin (Fig. 4.5). These cells are larger than other parenchyma cells in the persimmon fruit.

Looking at the components of condensed tannins (proanthocyanidin polymers) found in other plants, the polymers in apples and quinces from the Rosaceae family are made up of only catechins. While grapes are from the Vitaceae family, and bananas are from the Musaceae family, their components are catechins and gallocatechins (including a small proportion of gallate ester derivatives), and carob beans are plants from the Fabaceae family, but like persimmons, they are made up of four types of catechins (Aurore et al., 2023; Chira et al., 2009; Foo & Porter, 1980; Guyot et al., 2001; Liu & White, 2012; Lv et al., 2024; Matsuo & Ito, 1981a, b; Ohnishi-Kameyama et al., 1997; Shi et al., 2022; Souquet & Cheynier, 1996). In this comparison of chemical structures, the important points to note are whether the B ring has two or three phenolic hydroxyl groups, whether the C ring has a gallate ester bond (gallate), whether these catechins (components) are polymerized, and whether they are polymers, which seem to be major differences between each molecular species (Fig. 4.4). The details of the structural analysis of condensed tannins (kakitannin) in persimmons will be explained in Sect. 4.5.

4.2 Introduction to Kakishibu and Plastic—Traditional Society and Modern Society from the Perspective of Materials

This section compares traditional and modern societies through an examination of the chemical and physical properties of Kakishibu and plastics.

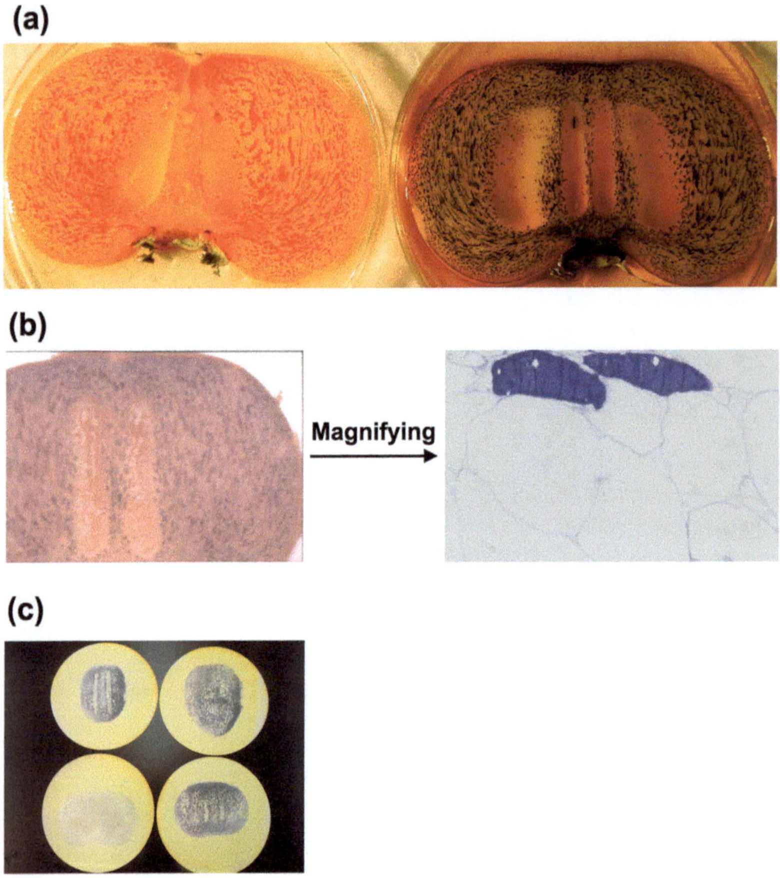

Fig. 4.5 Distribution of tannin cells scattered in the fruit of persimmon, stained with four methods (**a**): The fresh cut surface of the persimmon variety *Hiratanenashi* was stained with the hydrochloric acid-vanillin method (left) and the DMCAD (dimethyl cinnamaldehyde) method (right) to detect tannin cells. Staining method for proanthocyanidins (Abeynayake et al., 2011) (**b**): After the persimmon was treated to remove astringency, the cut surface was stained with ferric chloride to detect the distribution of insoluble tannin in the tannin cells (left). A portion of the fruit tissue embedded in epoxy resin was stained with methylene blue for microscopic observation of tannin cells (right). (**c**): After immersing in ferric chloride, the dried filter paper (tannin print) was pressed against the cut surface of the persimmon to confirm permeation of soluble tannin by color development. Tannin print procedure is cited from (Eaks, 1967; Matsuo et al., 1976). The soluble tannin that was insolubilized by the astringency removal treatment did not develop color (lower right). The remaining three were pressed against the fresh cut surface of the three different types of persimmons

4.2.1 Kakishibu as a Material and Traditional Society

Kakishibu made from immature green persimmons supported every aspect of traditional Japanese life. It is no exaggeration to say that it was used in all aspects of medicine, clothing, food, housing, daily necessities, and arts and crafts. There are few specialized documents on Kakishibu in traditional Japanese society, but here we introduce some valuable and comprehensive research results that have been carefully compiled through thorough field surveys in various parts of Japan, while touching on the history and customs of Kakishibu, especially manufacturing and application (Hirao, 2006, 2008; Imai, 2003, 2021; Yokoi et al., 1995). Traditional Japanese life was based on natural materials such as bamboo, *washi*, wood, cotton, and hemp, but they were prone to decay and lacked durability under Japan's hot and humid climate. Kakishibu, which gave them high water resistance and durability and made them less prone to decay, was the answer. The main component of Kakishibu is a natural condensed tannin, a polymer substance that is friendly to humans and the environment.

The reason why Kakishibu was so valuable was that it had various characteristics that other natural materials did not have. The astringent component, condensed tannin, in persimmons is initially colorless, odorless, and transparent in the fruit. Traditionally made Kakishibu, after fermentation and long-term aging, has a dark brown color with a reddish tint, like soy sauce. Kakishibu/condensed tannin in it usually shows the following characteristics:

1. Kakishibu has a strong astringency.
2. A high tannin-concentration liquid exhibits elevated viscosity at reduced temperatures and frequently undergoes gelation and solidification during cold and long storage.
3. Kakishibu/condensed tannin dissolves well in water and ethanol/methanol but does not dissolve in nearly 100% acetone or ethyl acetate. When the aqueous solution is dried, it forms a film and becomes film-like or solidifies and insoluble in both water and alcohols.
4. Kakishibu absorbs ultraviolet rays and emits weak fluorescence.
5. When heated in strong acid, Kakishibu produces red pigments called anthocyanidin.
6. When exposed to strong alkali in the presence of air, Kakishibu turns dark black-brown and decomposes.
7. Kakishibu/condensed tannin has strong antioxidant and radical scavenging effects. It prevents discoloration, fading, and odor generation in food. As oxidation progresses, it also functions as an oxidant.
8. Kakishibu/condensed tannin binds strongly with various proteins, causing aggregation and precipitation. This tanning action on raw hides, which is also the origin of the name tannin, was empirically known to have a strong binding ability with fibrous proteins. This binding and precipitation action of tannins with proteins was used practically in many situations.
9. Kakishibu inhibits the proliferation of viruses and microorganisms (Sakaguchi et al., 2014; Ueda et al., 2013). Traditional Kakishibu shows strong acidity, so one possibility is that this acidic solution inhibits the growth of microorganisms. Also, since condensed tannin in Kakishibu has a strong binding ability with a variety of proteins, it may bind to the surface proteins of microorganisms and inhibit their function and growth/proliferation. Recent studies have also reported that hydroxyl radicals and hydrogen peroxide are produced in a solution of polyphenols under certain conditions, resulting in bactericidal action

(Akagawa et al., 2003; Miyamoto, 2023; Nakamura et al., 2015). These possibilities need to be examined in more detail in the future to determine what is the main cause in each case.

10. Condensed tannin in Kakishibu has a strong binding ability with other HMW (high molecular weight) substances, such as polyvinyl alcohol, chitosan, chitin, starch, and methyl cellulose. When applied to *washi*, cloth or wood and dried, it forms a durable, water-resistant, and insect-resistant film while giving a brown color. Since ancient times, it has been traditional to apply Kakishibu to *washi* and dry it, which imparts a water-repellent property. Consequently, *bangasa*, traditional Japanese umbrellas and *kamiko*, paper clothes, were crafted and utilized daily during the Edo period. If the process of applying Kakishibu to *washi* and drying it is repeated many times, a hard film is formed and a certain degree of water repellency is observed.
11. Kakishibu/condensed tannin has a strong binding ability with alkaloids, e.g., caffeine, and a patent has been obtained for a method of removing caffeine from beverages by using Kakishibu.
12. Kakishibu/condesed tannin binds with various heavy metal ions such as cadmium and iron. In the field of dyeing, binding with iron ions is used to produce black to blue-black colors, with copper ions to produce red and with titanium ions to produce orange. In the field of environmental science, it is attracting attention for its use in the removal of heavy metals such as chromium, lead, and strontium ions and even in the recovery of precious metals.
13. Kakishibu/condesed tannin easily reacts and binds with various aldehydes.
14. Kakishibu/condesed tannin binds with volatile components that cause bad odors and exhibits a deodorizing effect.

4.2.2 Plastic as a Material and Modern Society

American John Wesley Hyatt invented celluloid in 1868. Celluloid is a semi-synthetic resin made by softening nitrocellulose with a plasticizer, which can be easily shaped when heated, and was highly promoted as a completely new material at the time. Later, Belgian chemist Leo Baekeland successfully developed phenolic resin, Bakelite, the world's first synthetic resin (plastic) by reacting phenol extracted from coal with formaldehyde in 1907 (Fig. 4.3). This resin, with its durability, chemical resistance, and high electrical insulation, was used in a wide range of applications, especially in industrial products (particularly aircraft and automobile parts) and electrical component housings. Subsequently, during World War II, as metal resources were prioritized for military use, the shortage of metals in the civilian sector became severe, and plastics were highlighted as an alternative to metals. The war accelerated technological innovation, and the development of new plastic materials was rapidly advanced.

The war greatly promoted the development of the plastic industry. Most plastics are made from naphtha, which is refined by heating and distilling crude oil (petroleum). Plastics are made from polymers that are chemically synthesized from naphtha. These plastic products, with their abundant and inexpensive raw materials, easy molding, free coloring, lightweight and durable finished products, and high thermal stability, were quickly used in various social situations. Especially in food-related containers, it is not an exaggeration to say that society has changed dramatically with the emergence of a variety of disposable products that can be molded into various shapes that are inexpensive, lightweight, and colorful.

4.2.3 Shift from Kakishibu Society to Plastic Society

Because of such technological and social changes, traditional natural materials like Kakishibu gradually moved to the margins of society as various plastic products from Western countries were quickly and widely introduced from the mid-Showa era (immediately after World War II). Compared to Kakishibu, these plastic products are extremely durable because they are artificially chemically synthesized, and they will not disappear by decomposing naturally or quickly when released into the environment. These properties of plastic have caused many environmental issues as well as human health problems caused by micro or nano plastics (Hasegawa & Kimura, 2022; Marfella et al., 2024; Parker et al., 2018; Worm et al., 2017).

Phenolic Resin and Condensed Tannin—Commonalities and Differences
Both phenolic resin and condensed tannin are HMW substances characterized by multiple benzene rings. However, phenolic resin is a polymer produced through chemical reactions involving LMW compounds obtained from coal. It constitutes a type of plastic that is inherently resistant to natural decomposition. On the other hand, condensed tannin is a polymer of polyphenols that plants synthesize and accumulate. It decomposes naturally by microorganisms and returns to the soil.

4.3 Scientific Analysis of Kakishibu Components

Traditional Kakishibu, produced by traditional methods, undergoes various chemical reactions and side reactions during fermentation, heat sterilization, aging, and storage, changing the components originally contained in the juice. Traditionally and coventionally, Kakishibu has been called by the following two names. Kakishibu, which is a light clay-yellow to light brown color with a slight greenish milky turbidity, fermented for less than a year after squeezing the astringent persimmon, is called *shinshibu* or *kishibu,* new and fresh Kakishibu. Once further aged and stored at room temperature in a dark place for 1–3 years, with the supernatant liquid scooped up after the sediment (dreg and turbidness) has sufficiently settled, it is called *koshibu* or *tamashibu*, old and aged Kakishibu. The color is normally dark black-brown and transparent. Thus, even though Kakishibu is referred to as a single term, there are various types, and therefore, their components differ depending on the kind of Kakishibu. Moreover, the components are expected to change during the aging process. This section focuses on the scientific analysis of the components of Kakishibu.

4.3.1 Traditional Way of Evaluating Kakishibu

Traditionally, Kakishibu artisans and *shibuya* have depended on their "intuition" to craft and utilize Kakishibu and Kakishibu products. However, for stable industrial use, we need to know more about its component. Additionally, quality standards are required for commercially sold Kakishibu, but in some cases, almost no data on the quality of Kakishibu is displayed. Sometimes, even the concentration of condensed tannin in it is not listed. Initially, the quality of Kakishibu was traditionally judged by the *toromi,* thickness, color, smell, and astringency when the craftsman scooped it up in a container that had been used for Kakishibu every year and dripped it from above. It was extremely sensory and is presumed to have changed under environmental conditions such as humidity and temperature. Quality control must have been quite difficult given seasonal changes. In traditional Kakishibu, the black-brown color and transparency were important, especially when used as dyes and paints, and it was the highest priority quality. Also, many *Shibuya*, traditional small-sized Kakishibu manufacturing and sales companies, similar to wineries, naturally blended some of Kakishibu to keep the quality constant. Later, in the *shibuya* that manufactured and sold Kakishibu, the Baumé degree (Bh) began to be used as a quality evaluation of Kakishibu for practical use. It is a so-called floating hydrometer that measures the *toromi* and concentration of Kakishibu by floating a long glass rod in the Kakishibu and reading the scale engraved on the glass wall to see how far it floats. However, it is heavily influenced by the viscosity of the liquid, which depends on temperature and impurities such as sugar. It has been valued for a long time and is still mainly used in specific locations.

After some time, a value called *rikika*, the titer of the colloidal titration method, began to be used for Kakishibu, that is intensively employed in the *orisage* process—a dreg- and turbidness-lowering treatment in the production of sake (Japanese rice wine), mirin (sweet sake as a seasoning), and vinegar. The *rikika* was a value that evaluated the protein precipitating ability of Kakishibu under certain condition and became a highly versatile and reliable standard method. However, the procedure was complicated and time-consuming, and it required a person who had acquired a certain level of chemical knowledge and technology. Also, it only evaluates its protein removal ability, which is one of the multifunctions of Kakishibu. Other scientific measurement approaches may be required when focusing on Kakishibu as a more global industrial material. Recently, among craftsmen who are making products using Kakishibu, there are sometimes voices saying that recent Kakishibu has changed, the color does not come out as before in the field of dyeing and painting of wood, and the adhesiveness of Kakishibu used between the layers of *washi* has deteriorated. What has changed within it so far?

4.3.2 Literature Reviews on Component Analysis of Kakishibu

In this section, I would like to review literatures on component analysis of Kakishibu.

4.3.2.1 Components in Different Kakishibu

In 1968 Nakabayashi conducted an analysis of four types of commercially available Kakishibu. They were made using the Japanese persimmon fruit (*Diospyros kaki*) as a raw material. In fresh juice, the Brix value, which is used to measure dissolved sugars, was 11 degrees, the tannin content was 3.53%, the protein removal ability using gelatin precipitation was 3.62, and the pH was 5.3. After fermentation of the fresh juice followed by aging (1–1.5 years), the values had changed to 6.3 ~ 7.8 degrees (Brix value), 4.74 ~ 5.96% (tannin content), 3.26 ~ 3.62 (protein removal ability), and 3.47 ~ 3.82 (pH) (Nakabayashi, 1968). These are fundamental data for comparing pre- and post-fermentation and aging of Kakishibu.

In addition, Akiyama et al. in 1969 collected Kakishibu from various parts of Japan, either commercially available or homemade, and analyzed their chemical components. They discovered that the color, Brix value, pH, acidity, and the tannin content varied in different Kakishibu. Especially, the tannin content, the functional component of Kakishibu, ranged from 0.75% to 5.75%. The higher the tannin content, the more effective Kakishibu is, so this result suggested that different Kakishibu had different effects. Regarding the production regions of Kakishibu, they pointed out that the Kakishibu circulating in the vicinity of Kyoto Prefecture had very high tannin content (Akiyama et al., 1969). This is because Kakishibu in Kyoto was mainly used for *orisage* of sake, which required high tannin content. On the other hand, in other regions than Kyoto, a lower content may have been used for other applications, such as dyeing *washi* and cotton cloth with Kakishibu. Furthermore, in 1971 Nakabayashi explained six types of commercially available Kakishibu in detail (Nakabayashi, 1971). In fresh juice of *Tenno* persimmon, the total acid was 0.29 (%), the total sugar was 7.84%, and the tannin content was 4.41% In these six different Kakishibu, compared to the fresh juice, the total acid increased to 1.79 ~ 2.48%, the total sugar decreased to 0.03 ~ 0.75, and the tannin content increased to 5.78% ~ 6.78%. Most of the total acid was acetic acid (52 ~ 69%), followed by propionic acid and butyric acid. This suggested that sugars were decomposed due to the alcohol fermentation in Kakishibu and then subsequent side reactions such as acetic acid fermentation produced multiple organic acids and changes in other components occurred. Regarding the increase of tannin content, it is presumed that evaporation of water caused concentration.

Finally, Imai et al. in 2001 conducted detailed component analysis of the aging process of Kakishibu manufactured in the laboratory using *Tenno* persimmon, famous as a raw material in the Minamiyamashiro region of Kyoto Prefecture, and *Tamura* persimmon, famous in Ibi District, Gifu Prefecture. Since commercially available Kakishibu was often made by mixing raw material persimmons, the authors were trying to understand temporal changes during the manufacture of

Kakishibu more accurately under a non-mixed condition. They discovered that fresh juice of *Tenno* persimmon had higher soluble tannin content, Brix value, and colloid titration value than the fresh juice of *Tamura* persimmon. They reported that fermentation occurred vigorously from 3 to 29 days, and intense foaming was observed during the subsequent aging period. Total sugar decreased sharply on the eighth day after the preparation of Kakishibu and could not be detected thereafter. Along with this, various organic acids increased, especially acetic acid, which increased rapidly from 15 to 29 days. The soluble tannin content gradually decreased during the long-aging period. About 6 months later, Kakishibu derived from *Tenno* persimmon had more tannin content and less organic acid than that from *Tamura* persimmon. This was the first and only detailed observation of single-species-origin Kakishibu making process (Imai et al., 2001).

4.3.2.2 Odor Components of Kakishibu and Techniques to Reduce Them

In recent years, the demand for Kakishibu that reduces the generation of irritating and putrid odors has increased.

Yoshida et al. described in a pioneering report that using a special yeast isolated and selected from *Tenno* persimmon could reduce unique irritating and putrid odors (Yoshida et al., 1985).

Furthermore, Iwamoto et al. in 2013 reported that ultrafiltration could only remove foul odor components, while not removing tannin, the functional component of Kakishibu. This ultrafiltration is an important technology because it can minimize the influence of LMW (low molecular weight) substances that do not significantly contribute to the known characteristics of Kakishibu (Iwamoto et al., 2013). However, in this ultrafiltration technology, it is important to select the material of the ultrafiltration membrane appropriately because HMW tannin strongly binds with many material substances.

4.3.3 Easier and Effective Ways of Evaluating Kakishibu

It has been represented based on Brix and Baumé degree, which are simple indicators, but these indicators are easily affected by other impurities and do not accurately represent tannin content. Therefore, we decided to compare the traditional Brix value, Baumé scale (Bh), and rikika of colloid titration titer with other modern analytical values. Knowing the correlation between the values of traditional indicators and the values of current chemical analyses is essential for examining the effectiveness of simple indicators. If there is a way to easily grasp the quality of Kakishibu, it can greatly contribute to the development of the Kakishibu industry.

In order to undertake these comparisons, I purchased two types of commercially available Kakishibu and applied various old and new analysis methods to reexamine the quality of Kakishibu. In this experiment, the same sample was dialyzed before and after the analysis with the aim of eliminating the influence of LMW compounds involved in the analysis as much as possible. Also, to gauge the effect of aging, *shinshibu,* new and fresh Kakishibu, and *koshibu,* old and aged Kakishibu, were compared.

In particular, I attempted to find out a method that can be easily and accurately carried out. As a quality evaluation that can be analyzed quickly with a simple device, we measured the dry matter weight ratio (%) of the sample using a heated moisture meter. The condensed tannin in Kakishibu was quantified using three types of colorimetric quantification (Folin-Ciocalteu method, vanillin-sulfuric acid method, and DMCAD method) commonly used to measure the amount of polyphenols in various foods and plant raw materials, and each quantification value was compared and examined (Cáceres-Mella et al., 2013; Neilson et al., 2016; Schofied et al., 2001; Shi et al., 2021). However, at present, no matter what accurate analysis method is used, there is no proper standard substance for the astringent component of persimmon, condensed tannin, so there is a limit to the accuracy of tannin quantification. Rather than scientific accuracy, I would like to instead discuss how to estimate condensed tannin in Kakishibu in a simple way. The results are shown in Table 4.1.

The *shinshibu* analyzed here was a milky, highly viscous substance (solution). The *koshibu* was transparent, dark black-brown, and slightly sticky. During the analysis, the *shinshibu* gradually formed continuous precipitates and films on the

Table 4.1 Physical and chemical analysis of two types of Kakishibu (new and old) commercially available in Kyoto Prefecture, conducted by the author

		Samples			
		Shinshibu (new fresh Kakishibu)	*Dialyzed*[a] *Shinshibu* (new fresh Kakishibu)	*Koshibu* (old aged Kakishibu)	*Dialyzed*[a] *Koshibu* (old aged Kakishibu)
Analysis items	pH	3.25	3.33	3.36	3.4
	Baumé value	5.9	2.9	4.7	3.2
	Viscosity	3.58	2.75	2.73	2.28
	Dry matter weight (%)	9.8	4.7	8.6	5.4
	1. Tannin content by Folin-Ciocalteu method	3.97	3.53	4.46	4.25
	2. Tannin content by vanillin-sulfuric acid method	1.98	1.6	1.97	1.5
	3. Tannin content by DMCAD method	0.52	0.42	0.58	0.38

[a]Dialyzed with a membrane (a chemically modified cellulose membrane; visking tubing for dialysis) with a molecular weight cut-off profile of <14,000 to remove substantially LMW impurities in the same Kakishibu sample. The rikika values were provided by the Kakshibu manufacture; 148 for the shinshibu and 163 for the koshibu.

surface of the liquid. The differences between the two can be seen in the viscosity and Baumé (Bh) degrees (3.58 and 5.9 in *shinshibu* and 2.73 and 4.7 of *koshibu*). After removing LMW substances by dialysis, both values significantly decreased (3.48 to 2.75 of viscosity and 5.9 to 2.9 of Bh in *shinshibu,* and 2.73 to 2.28 of viscosity and 4.7 to 3.2 of Bh in *koshibu*), indicating that the Bh degrees were greatly affected by turbidity, colloidal substances, and LMW substances. The dry matter weight (%) more clearly demonstrated the effect of removing LMW substances (9.8% to 4.7% in *shinshibu* and 8.6% to 5.4% in *koshibu*). The larger decrease observed in *shinshibu* than *koshibu* is likely due to the presence of fine particles produced during fruit crushing and squeezing, which cause turbidity.

Regarding the dry matter weight after dialysis, they were thought to be mainly condensed tannins (HMW substances). In fact, the tannin amount calculated by the Folin-Ciocalteu method in terms of epigallocatechin was approximated to the dry matter weight. Since there is a 2–3% difference in dry matter weight before and after dialysis, it is thought that a part of LMW compounds also remained.

Next, compared to the results of the Folin-Ciocalteu method (3.97 in *shinshibu* and 4.46 in *koshibu*), the other two types of analysis methods (vanillin-sulfuric acid method and DMCAD method) showed completely different low values (1.98 and 0.52 in *shinshibu* and 1.97 and 0.58 in *koshibu*). This is because of the completely different reaction principles shaping the color development of the three. Regarding the Folin-Ciocalteu method, the color appeared by the reducing action of polyphenols. The reaction mode of DMCAD and vanillin-sulfuric acid with proanthocyanidin under strong acidic conditions is not the same. The DMCAD is reported to specifically react with the terminal flavan-3-ol unit of proanthocyanidin, while vanillin-sulfuric acid is known to react not only with the terminal unit but also with the extension units in it. Probably, considering the reaction mode of these two, it is thought that the reaction with condensed tannins occurs only with a part of the whole molecule, and therefore, the color development became low (Hagerman & Butler, 1994; McMurrough & McDowell, 1978; Oszmianski & Bourzeix, 1996; Shi et al., 2021).

Finally, in that the dry matter weight, providing values close to true tannin content, was 1.3 times higher in *shinshibu* and1.3 times in *koshibu* than the tannin amount estimated by the Folin-Ciocalteu method after dialysis. Folin-Ciocalteu method (a colorimetric method) could be one of good indicators of the tannin amount. The following can be inferred about the discrepancy in the quantification values between dry matter weight and estimated tannin amount by Folin-Ciocalteu method (Table 4.1):

1. The setting of the standard substance for this quantification method was inappropriate.
2. There is a limit to the reaction of the Folin-Ciocalteu method to tannin.
3. The influence of LMW compounds.
4. The decomposition of tannin during dialysis, or the loss of tannin due to adsorption to the dialysis membrane.

For easy implementation of this method, I suggest using a commercially available water quality test kit to measure phenolic substances, which allows even people without chemical skills to check roughly. Also, without using any reagents or equipment, and without the need for complicated procedures, a simple quantitative comparison of condensed tannin using a tannin print (a filter paper impregnated with ferric chloride and dried) that looks at the binding with iron ions seems to be usable on site (Eaks, 1967; Matsuo et al., 1976, 1984)

4.4 Science of the Chemical Structure and Molecular Weight of Kakitannin

The main component of Kakishibu (a type of condensed tannin, called kakitannin), was a fairly familiar natural material in Japan, but its chemical entity was unknown for a long time. Even in the world, research on the structure of tannin had hardly not progress for a long time. The hot water extract of bark and heartwood, which has been used as a raw hide tanning agent since ancient times, was called tannin, but its chemical substances and structures remained unknown. This extract was astringent, and then, astringent things were often simply and conventionally called tannin.

In this section, I will review the science on the chemical substances and structures of kakitannin.

Kakitannin, Tannin Contained in Kakishibu

Kakishibu has a strong astringency, so it contains HMW tannins (a type of condensed tannin) at high concentrations. But it is fermented and aged, and the tannin that existed in the original fruit and the tannin that changed due to various factors during fermentation, aging, and heat sterilization are mixed. This ratio changes over time. Therefore, I decided to call the unchanged ones in the fruit "kakitannin," condensed tannin of persimmon, and the changed ones during fermentation and aging "changed kakitannin" to distinguish them. I believe that most of the solution immediately after squeezing exists as kakitannin. Kakitannin itself is almost transparent, colorless, and odorless, but changed kakitannin is light reddish-brown to dark black-brown, with a soy sauce-like color. The changed kakitannin gradually grows into a large colloid during aging and storage and settles as dreg and turbidness. After long-aging the old Kakishibu is usually transparent and dark black-brown.

4.4.1 The Dawn of Kakitannin Scientific Research

The chemical research on the astringent components of astringent persimmons, kakitannin, in Japan was essentially first done by Komatsu et al. of Kyoto University in 1923 (Komatsu & Matsunami, 1924, 1925; Komatsu et al., 1925). At that time, there was no suitable isolation technique or method to analyze the structure, so ultimately, he named the crude substance of astringent persimmon "shibuol" and proposed its chemical formula. He only recognized the presence of gallic acid and phloroglucinol in the molecule. After a long hiatus in research, Ito and Oshima in 1962 found that when the precipitate obtained by adding glacial acetic acid to the methanol extract of astringent persimmon was heated under high concentration hydrochloric acid, a type of red pigment, delphinidin, and glucose were produced. Then, Ito and Joslyn in 1964 confirmed that the HMW tannins of astringent persimmon were prepared by two methods, the glacial acetic acid precipitation method and the lead acetate precipitation method; both produced a small amount of gallic acid, gallocatechin, and gallocatechin gallate by hydrolysis with weak acid. They speculated that the main body of kakitannin was a conjugated tannin (Ito & Joslyn, 1964; Matsuo & Ito, 1977; Zhang et al., 2011). However, the veil of the mystery of its reality remained and many years passed.

4.4.2 Hints from Tannin Research of Other Plants

The key to solving the mystery of this astringent substance of persimmons was brought by polyphenol research of other plants. Weinges et al. (1969) introduced the findings that there was a substance in the leaves of young grapes that produced a type of red pigment of flowers, anthocyanidin, by strong acid treatment, and the presence of a substance in cocoa beans that produced cyanidin by acid treatment in their review article. In it they proposed that there were two types of compounds that produced red pigments by strong acid treatment, extracted from plants, and they called them leucoanthocyanidin and condensed proanthocyanidin (Haslam, 1975; Weinges et al., 1969; Weinges, 1975).

4.4.3 Discovery of the Chemical Structure of Kakitannin

In 1975, the author and Ito carefully examined the research results on polyphenols and tannins that were expected to be HMW substances extracted from other plant species until then and started a new challenge. First, we examined various methods to obtain high-purity HMW polyphenolic substances by extracting water-soluble or alcohol-soluble substances from astringent persimmon juice under conditions where decomposition or denaturation does not occur as much as possible. Eventually, we prepared high-purity tannin from the methanol extract of immature fruit of astringent persimmon (*Hiratanenashi* persimmon) by the disodium hydrogen phosphate precipitation method and analyzed its chemical structure (Matsuo & Ito, 1978). When this HMW tannin preparation was decomposed with 2 N hydrochloric acid or sulfuric acid, the reaction solution turned red-purple, and two types of red pigments, cyanidin and delphinidin, could be detected in a ratio of about 1:3. Furthermore, we analyzed the HMW tannin by referencing the method used by

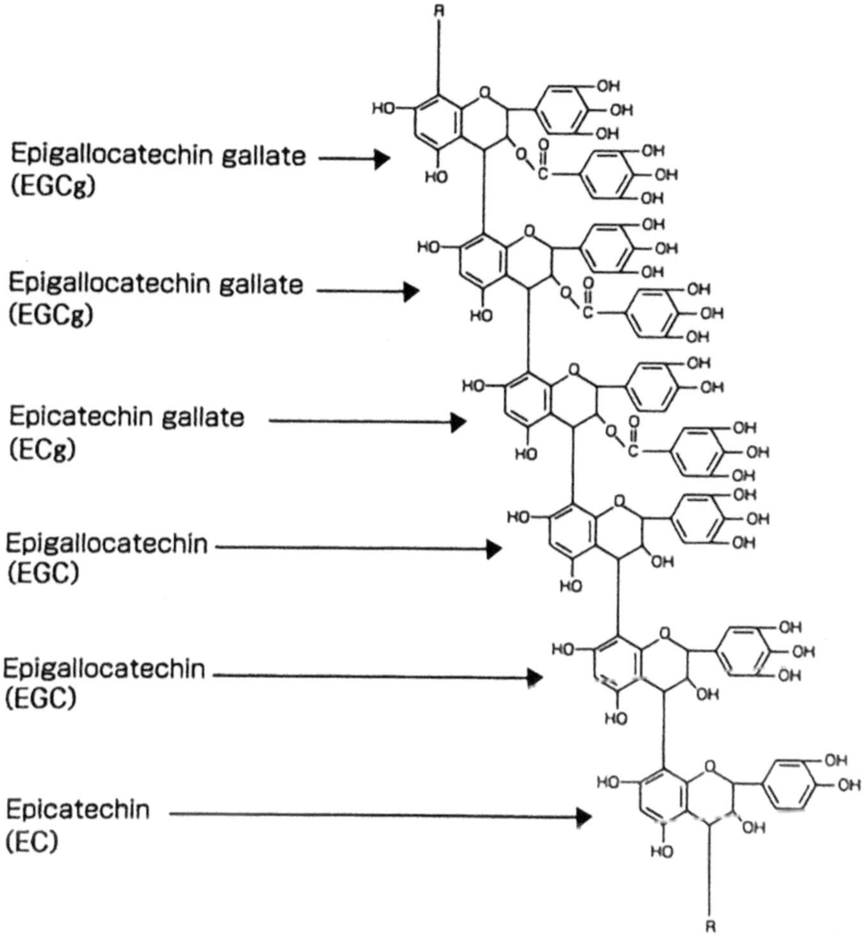

Fig. 4.6 The chemical structure of kakitannin

Thompson et al. in 1972 for the structural analysis of the proanthocyanidin B group (dimer), specifically the decomposition reaction with a thiol compound (a nucleophilic reagent) in acidification (i.e., thiolysis) (Thompson et al., 1972). As a result, I and Ito in 1978 succeeded in decomposing it into four types of constituents very efficiently, which could hardly obtain decomposition products by acid hydrolysis as the HMW tannin until then. When we isolated these constituents and analyzed their chemical structures by instrumental analysis, we found that thiol derivatives of catechin, catechin gallate, gallocatechin, and gallocatechin gallate were produced in a ratio of 1:2:1:2 (Matsuo & Ito, 1978). We finally inferred that the B-type proanthocyanidin polymer being catenated these four types of catechins was the entity of kakitannin (Fig. 4.6).

4.4.4 Attempts at Molecular Weight Determination of Kakitannin

Next, we tried to get a clue about its molecular weight using GPC (gel permeation chromatography). At that time, since suitable column carriers, solvents, and standard substances for analyzing water-soluble HMW compounds were not easily available, we chemically modified the many hydroxyl groups of kakitannin to prepare a methylated derivative as a sample. We used THF (tetrahydrofuran) as a solvent and analyzed the molecular weight of the derivative using polystyrenes of various molecular weights as standard substances. As a result, $\overline{Mw} = 1.38 \times 10^4$ (weight average molecular weight) and $\overline{Mn} = 0.64 \times 10^4$ (the number average molecular weight) were obtained for the methylated kakitannin. At that time, we had not fully considered the degradation and side reactions of the proanthocyanidin polymer during chemical modification, so it was unclear how much this molecular weight value reflected the true value of the unmodified kakitannin (an intact condensed tannin of persimmon).

Other analysis of HMW tannins by GPC using organic solvents after chemical modification was attempted, but there were problems with changes in the sample during the chemical modification reaction, the type of appropriate solvent and column carrier (the packing material particles) for GPC, and the presence of appropriate molecular weight standards, and satisfactory results cannot be said to have been obtained (Wu & Hwang, 2002; Stringano, et al., 2011; Shi, et al., 2021). Differential refractometer (RI)-GPC analysis is usually a conversion value based on comparison with standard synthetic polymers such as polystyrene, so it seems impossible to determine the true molecular weight (absolute molecular weight) of polymers with different chemical structures. In recent years, analysis by size exclusion chromatography (SEC) of proanthocyanidin polymers has also been studied (Stringano et al., 2011; Yanagida et al., 2002).

The author (Matsuo) conducted a simple preliminary test to find a suitable solvent for the analysis of kakitannin. Cubes of naturally solidified Kakishibu (5 mm cubes) were placed in various solvents, and those that could be dissolved within 1 h by stirring were selected. As a result, it was found that DMSO (dimethyl sulfoxide) and DMF (dimethylformamide) could dissolve most effectively.

Therefore, it was found that the better solvent for SEC-HPLC analysis was one of their solvents with 2 mM LiBr added (Matsuo, unpublished). Currently, we are considering a column carrier with a wide fractionation range that can properly analyze intact kakitannins. Compared to traditional methods, analysis by chromatography equipped with a multi-angle laser light scattering detector called GPC/SEC-MALS can provide information on the absolute molecular weight, molecular weight distribution, gyration radius, and degree of branching of organic polymers. This system is said to be able to analyze HMW substances in the range of absolute molecular weight from several thousand to about 30 million and in the range of about 10 nm to several hundred nm. By finding the appropriate solvent and column and analyzing intact kakitannins with this system, we expect to gain completely new insights. Historically, organic solvent-based GPC and aqueous SEC were distinguished, but today they are often considered the same analytical technique since they share the same separation mode.

As another approach, Nunokawa and Mena in 1973 have been studying the role of Kakishibu used for *orisage*, a dreg- and turbidness-lowering treatment in the production of sake, mirin (sweet sake) and vinegar, for many years. In the process, they used the ultracentrifugal sedimentation method to measure the particle size of Kakishibu, tannic acid, protein turbid substances, and tannin-protein turbid substance complexes. Using an analytical ultracentrifuge, they centrifuged and sedimented each particle at 55,000 ~ 10,000 rpm and analyzed them. As a result, they found that these sizes were 12 ~ 14 mμ, 4 mμ, 13 mμ, 140 mμ, 250 ~ 400 mμ. It is extremely interesting to focus on the size of each particle when considering the protein removal mechanism of Kakishibu. The Kakishibu used here was 3 years after production, and it is unclear what changes occurred during fermentation, heat sterilization, aging, and storage. The sake was also an old sake that had been aged for about a year (Nunokawa & Mena, 1973).

The author (Matsuo) extracted and purified the intact kakitannin under careful conditions that did not denature as much as possible, and requested the measurement of static light scattering using Dynamic Light Scattering Photometer DLS-8000(Otsuka Electronics, Japan) after replacing the solvent of kakitannin samples to MeOH or DMF by using a dialysis procedure. As a result of this analysis, it was shown that kakitannin had a large molecular size exceeding 120 nanometers and a molecular weight of over one million daltons (Matsuo, unpublished). I consider that these results require multiple retests with higher reproducibility, and the question is whether intact kakitannin molecules exist as single molecules during analysis without associating or polymerizing, or conversely, whether they have not decomposed during preparation. More careful analysis is awaited in the future.

4.4.5 *Successive Reports on the Chemical Structure Determination of Kakitannin*

Piretti et al. in Italy reported that the ratio of the constituent components of the proanthocyanidins of astringent persimmons was 2:3:3:4, and they speculated that the slight difference in this component ratio from the results of Matsuo and Ito (1978) was due to the difference in the variety of astringent persimmons (Piretti et al., 1985). Subsequently, other reports also pointed out the presence of other polyhydric HMW compounds such as pectin bound to tannin, but it is believed that there are still issues to be reconsidered such as the preparation method and purity of kakitannin.

In 1994, Tanaka of Nagasaki University in Japan reported the results of a rapid and accurate analysis of the degradation products produced by thiolysis of the tannin substances (proanthocyanidin polymers) of astringent persimmons by HPLC (Tanaka et al., 1994). This method has been shown to be applicable to insolubilized tannin in artificially astringency-removed fruit. This was a valuable insight in understanding the mechanism of astringency reduction during the maturation of astringent persimmons and the mechanism of artificial de-astringency technology. Gu et al. (2008) found that the distribution of the tannins extracted with acidic methanol from persimmons ranged from 1.16×10^{4} Da to 1.54×10^{4} Da, with a number average molecular weight (Mn) of 1.28×10^{4} Da and a weight average molecular weight (Mw) of 1.39×10^{4} Da. They also reported that the ratio of the constituent components of the HMW fraction analyzed by thiolysis was 1:3:1:7 for unknown monomer: epicatechin gallate (ECg): epigallocatechin (EGC): epigallocatechin gallate (EGCg) (Gu et al., 2008). In 2010, Li, C. who joined the group of Hagerman, A.E. in the United States, analyzed in detail the tannins prepared from mature persimmon fruit (*Diospyros kaki* L., cv. Shanxi) harvested in China at the end of October. The results were quite different from those obtained so far. While EC, ECg, EGC, EGCg (1,3,1,4) were similar to the previous results as the constituent components of the proanthocyanidin extension chain, they announced the presence of proanthocyanidins with A-type bonds, in addition to B-type bonds, with the terminal components being catechin, ECg, and further myricetin (Li et al., 2010). (EC is an abbreviation for epicatechin) They also reported that the average degree of polymerization was 19 to 47, which determined by thiolysis-HPLC analysis. While, the calculated Mn and Mw from GPC were 50% to 70% lower than these results. At the same time, Kawakami et al. (2010) reported that in the analysis of HMW tannins derived from the leaves of Japanese astringent persimmon varieties such as cv. *Hiratanenashi* and *Tonewase*, the ratio of the constituent components was EC:ECg:EGC:EGCg = 7:2:8:1, and the terminal unit was (+)-catechin, indicating that the structure of the proanthocyanidin polymer contained in the leaves was quite different from that in the fruit (Kawakami et al., 2010).

Subsequently, detailed research on the tannin substances of astringent persimmons was conducted by Yonetani at Kinki University. From the analysis of decomposition products, it was announced that Kakishibu tannin was a proanthocyanidin polymers composed of EC:ECg:EGC:EGCg = 1:1:3:2 components, and that the terminal component (unit) of its molecular chain was ECg (Takemori et al., 2016). Yonetani and his team members further applied similar methods to many persimmon varieties and analyzed the differences in their components in detail (Takemori et al., 2022). Their analysis samples (Kakishibu) used a special preparation method called Nara-style Kakishibu, which may have been different from other traditional Kakishibu (Hamazaki, 2010). The difference in the components of this kakitannin or Kakishibu is expected to affect the physical, chemical, and physiological properties of these tannins, and more precise research is desired in the future. However, it is speculated that the tannin content and the molecular size of the tannin in the astringent persimmons have a stronger influence on the characteristics of Kakishibu.

Other studies on proanthocyanidins are very interesting and valuable, such as the study of grape astringency components related to wine flavor research (Liu & White, 2012; Rodriguez-Perez et al., 2019) and the study of astringency components that started from the effective use of young fruits of thinned apples (Guyot et al., 2001). However, it is expected that the molecular size of kakitannin is much larger compared to the previous two. Therefore, significant differences are observed in some physical properties, such as gelation at low temperatures, high viscosity, and film formation when dried.

4.4.6 Challenges and Prospects on Extraction and Decomposition of Kakitannin

In the analysis of the components of kakitannin, it is necessary to consider the differences in various astringent persimmon varieties and cultivation areas, but the most important point is the method and process of extracting the tannin and the conditions of storage. In particular, whether the tannin has been precipitated, dried, or exposed to harsh conditions such as heating or strong alkali/strong acid. They can cause the tannin to incorporate other impurities, decompose, bind secondarily, and various side reactions may occur. Careful attention is required for sample preparation and storage in the structural analysis of kakitannin, HMW proanthocyanidin. In other words, it is considered essential to analyze using the tannin preparations that have been extracted and purified under conditions that do not denature the tannin in the fruit as much as possible. On the other hand, Kakishibu (especially, *koshibu* and *tamashibu,* old aged Kakishibu) are fermented, sterilized, long aged, and stored, the tannins present in them are expected to have undergone various denaturation, and it is thought to be a mixture of denatured and changed kakitannin with different degrees of aging and denaturation. In contrast to the intact kakitannin in the fruit, or the tannin immediately after squeezing the astringent persimmon, it is structurally very different, and at present, scientific analysis of denatured and changed kakitannin is thought to be difficult. To put it simply, it is like comparing the polyphenols of green tea with the fermented, aging, and denatured polyphenols of black tea. Many types of denatured polyphenols not found in green tea have been discovered in black tea (Hashimoto et al., 1992; Haslam, 2003; Ngure et al., 2009; Tanaka & Koumoto, 2003). Considering that it has been clarified that the astringent components of astringent persimmons (kakitannin) have the structure of proanthocyanidin polymers, the following chemical properties are inferred, and they were sufficient to fully explain the characteristics of Kakishibu (mainly kakitannin, mixed with denatured and changed one) that have been described for a long time.

1. It has many phenolic hydroxyl groups in the molecule; therefore, it shows strong reducing action and is a weakly acidic substance. Air oxidation progresses rapidly under weakly alkaline to alkaline conditions and results in darkening. It forms many hydrogen bonds with many substances and binds strongly. Under

high concentration and low temperature, it easily forms intra- and intermolecular bonds, becomes a high viscosity solution, and eventually gels.

2. It has two or three adjacent phenolic hydroxyl groups: It easily binds with most of heavy metal ions and produces their unique colors.
3. It has many benzene rings in the molecule: It shows partial lipophilicity and forms hydrophobic bonds and physical bonds. It has properties intermediate between hydrophilicity and lipophilicity.
4. It is a polymer (HMW compound) made up of four types of catechins: Because it forms many hydrogen bonds and hydrophobic bonds within and between molecules, it can become a high viscosity liquid under certain conditions, and when applied to *washi* or wood, it forms a strong film after drying, increasing water repellency, water resistance, insect resistance, and durability.

As previously mentioned, this thiolysis (thiol degradation) method greatly opened the way to chemically elucidate unknown complex substances that had been vaguely defined as condensed tannins or HMW tannins. In other words, it was found that they could be properly decomposed into constituent components at a high yield. The reaction efficiency of this thiolysis is also being studied (Guyot et al., 2001; Matthews et al., 1997). It has been reported that in the homopolymer of apple's (−)-epicatechin, the decomposition yield is about 75% and this yield is not related to the degree of polymerization. However, although degradation by thiolysis was an effective method, there was a strong foul odor due to the residual sulfur-containing reagent when examining the physiological effects of the degradation products, which was a problem. Therefore, when we also examined the degradation of kakitannin by phloroglucinol, which was expected to act as a nucleophilic reagent, we found that it significantly promoted the degradation of kakitannin under acidic conditions (Akagi et al., 2010; Kennedy & Jones, 2001; Matsuo et al., 1984; Matthews et al., 1997). A comparison of the thiolysis method with decomposition by phloroglucinol has been studied, and its superiority has been reported (Kennedy & Jones, 2001; Matthews et al., 1997).

4.4.7 Big Remaining Issues on Kakitannin

The big remaining issues regarding kakitannin (proanthocyanidin polymer), the actual substance of persimmon astringency, are to clarify the overall arrangement of its catechin derivatives and the overall three-dimensional structure and dynamic behavior of whole molecule in solution, and the exact molecular size. So far, the components of kakitannin and their ratios, as well as the identification of terminal molecules, have been reported in several papers as already mentioned. However, there is currently no way to study the arrangement and order of these components. If a method can be developed to cut them in order from the end, it may be possible to clarify. Another approach is to change in steps the amount of thiol compound added and the acid concentration to perform partial thiolysis and then connect the huge number of fragment sequences obtained by fragment analysis with computer processing.

Roplak et al. in 2017 revealed that in condensed tannin-protein aggregation tannin size (mDP; mean degree of polymerization or amw; average molecular weight) was the most important parameter, not the composition of the polymer (Roplak et al., 2017). Recently, Yamauchi et al. in 2023 have analyzed the structure of relatively LMW proanthocyanidin oligomers and then estimated the binding position of tannin and protein by NMR method and docking simulation. Furthermore, they have created and published a model of their aggregation (Yamauchi et al., 2023). In the future, it is highly likely that completely new insights will be found if these research methods are applied to HMW proanthocyanidin polymers, such as kakitannin.

Regarding the molecular size, if my current estimation of over one million daltons (Matsuo, unpublished) are close to the intact molecular size, it raises further challenges. That is, it is a very interesting issue how the single molecular chain of kakitannin takes what molecular form in each solution, how the result affects the physical properties, and how it affects the physiological action. Also, it is about the binding mode of kakitannin and protein in solution and the mechanism of precipitation. The molecular form of this kakitannin may also affect the quantitative values by chemical color reaction. Furthermore, it seems that the mystery is digging up the next mystery, such as how kakitannin is synthesized during biosynthesis in tannin cells in the fruit and accumulated at high concentrations while remaining soluble in water.

References

Abeynayake, S. W., Panter, Mouradov, A., & Spangenberg, G. (2011). A high-resolution method for the localization of proanthocyanidins in plant tissues. *Plant Methods, 7*, 13–19.

Akagawa, M., Shigemitsu, T., & Suyama, K. (2003). Production of hydrogen peroxide by polyphenols and polyphenol-rich beverages under quasi-physiological conditions. *Bioscience, Biotechnology, and Biochemistry, 67*(12), 2632–2640.

Akagi, T., Suzuki, Y., Ikegami, A., Kamitakahara, H., Takano, T., Nakatsubo, F., & Yonemori, K. (2010). Condensed tannin composition analysis in persimmon (*Diospyros kaki* Thunb.) fruit by acid catalysis in the presence of excess phloroglucinol. *Journal of the Japanese Society for Horticultural Science, 79*(3), 275–281.

Akiyama, H., Uchiyama, K., Koide, I., Himeno, K., & Noda, Y. (1969). Seishu seicho-yō Kakishibu no hinshitsu ni tsuite (Daiippō) Kakishibu no seibun ni tsuite (Quality of kakishibu for purifying sake (first report): Components of kakishibu). *Journal of the Society of Brewing, Japan, 64*(10), 889–893.

Appel, H. M. (1993). Phenolics in ecological interactions: The importance of oxidation. *Journal of Chemical Ecology, 19*(7), 1521–1552.

Aurore, G., Vingadassalon, A., Bourvellec, C. L., Dufour, C. B., & Renard, C. M. G. C. (2023). Presence of highly polymerized proanthocyanidins in pulp and peel of unripe bananas (*Musa* sp.) from the French West Indies. *Journal of Food Technology and Preservation, 5*(3), 145–152.

Bajec, M. R., & Pickering, G. J. (2008). Astringency: Mechanisms and perception. *Critical Reviews in Food Science and Nutrition, 48*(9), 858–875.

Barbehenn, R. V., & Constabel, C. P. (2011). Tannins in plant–herbivore interactions. *Phytochemistry, 72*(13), 1551–1565.

Cáceres-Mella, A., Peña-Neira, A., Narváez-Bastias, J., Jara-Campos, C., López-Solís, R., & Canals, J. M. (2013). Comparison of analytical methods for measuring proanthocyanidins in wines and their relationship with perceived astringency. *International Journal of Food Science & Technology, 48, 2588–2594.*

Chira, K., Schmauch, G., Saucier, C., Fabre, S., & Teissedre, P.-L. (2009). Grape variety effect on proanthocyanidin composition and sensory perception of skin and seed tannin extracts from Bordeaux wine grapes (Cabernet Sauvignon and Merlot) for two consecutive vintages (2006 and 2007). *Journal of Agricultural and Food Chemistry, 57*(2), 545–553.

Eaks, I. L. (1967). Ripening and astringency removal in persimmon fruits. *Proceedings of the American Society for Horticultural Science, 91*, 868–875.

Foo, L. Y., & Porter, L. J. (1980). The phytochemistry of proanthocyanidin polymers. *Phytochemistry, 19*(8), 1747–1754.

Gu, H.-F., Li, C.-M., Xu, Y.-J., Hu, W.-F., Chen, M.-H., & Wan, Q.-H. (2008). Structure features and antioxidant activity of tannin from persimmon pulp. *Food Research International, 41*(2), 208–217.

Guyot, S., Marnet, N., & Drilleau, J. (2001). Thiolysis-HPLC characterization of apple procyanidins covering a large range of polymerization states. *Journal of Agricultural and Food Chemistry, 49*(19), 14–20.

Hagerman, A. E., & Butler, L. G. (1994). Assay of condensed tannins or flavonoid oligomers and related flavonoids in plants. *Methods in Enzymology, 234*, 429–437.

Hamazaki, S. (2010). Ethanol de datsuju shita kajitsu o mochiita kakitannin no jinsokuna chōsei-hō(Quick preparation of kaki-tannin from de-astringent persimmon fruit treated by ethanol). *Horticultural Research in Japan, 9*(3), 367–372.

Hasegawa, K., & Kimura, Y. (2022). Kaiyō microplastic osenmondai no genjō (Present situation and problems of marine microplastic pollution). *Report of the Marine Ecology Research Institute, 27*, 11–20.

Hashimoto, F., Nonaka, G., & Nishioka, I. (1992). Tannins and related compounds. CXIV. Structure of novel fermentation products, theogallinin, theaflavonin and desgalloyl theaflavonin from black tea, and changes of tea leaf polyphenols during fermentation. *Chemical and Pharmaceutical Bulletin, 40*, 1383–1389.

Haslam, E. (1975). The chemistry and biochemistry of plant proanthocyanidins. In L. Farkas (Ed.), *Topics in flavonoid chemistry and biochemistry*. (Proceedings of 4th Hungarian Bioflavonoid Symposium (pp. 77–99). Elsevier Science.

Haslam, E. (1979). Vegetable tannins. In T. Swain, J.B. Harborne, & V. Sumere (Eds,), *Biochemistry of plant phenolics* (pp. 475–523). N.Y. and London: Plenum press.

Haslam, E. (2003). Thoughts on thearubigins. *Phytochemistry, 64*, 61–73.

Harborne, J. B. (1977). *Introduction to ecological biochemistry*. Academic Press, London.

Hirao, K. (2006). Kakishibu riyō bunka no rekishi to shokuhin kakō e no dōnyū ni kansuru kiso kenkyū (The basic research on the history and culture of use of persimmon juice, and the induction to food processing). *Asahi Beer Foundation Research Report*, 87–96.

Hirao, K. (2008). Nihon no gendai shokuseikatsu ni miru Kakishibu no katsuyō-hō (Utilization of Kakishibu in modern Japanese food life). *Kaishi Shokubunka Kenkyu (Journal of Food Culture Research), 4*, 25–32.

Horii, N. (2006). *Uji cha o kataritsugu (Passing on Uji tea)*. Earth.

Imai, K. (2003). *Kakishibu: Mono to Ningen no Bunkashi 115 (Kakishibu: Cultural history of things and humans)*. Hosei University Press.

Imai, K. (2021). *Kaki: Mono to Ningen no Bunkashi 185 (Persimmon; cultural history of things and humans)*. Hosei University Press.

Imai, K., Shiozaki, S., Tsuda, T., Nakao, Y., Ogata, T., & Horiuchi, S. (2001). Kakishibu no jukusei katei ni okeru tō, yūkisan, oyobi 2, 3 no tokusei no henka (Changes in sugar, organic acid and characteristics of kaki-shibu during fermentation). *Journal of the Japanese Society for Horticultural Science, 70*(1), 95–101.

Ishigaki, K. (1981). Ocha no kagaku seibun: Aji, kaori to chaju no saibai (Chemical components of tea: Taste, aroma, and tea cultivation). *Kagaku to Seibutsu (Chemistry and Biology), 19*(5), 278–285.

Ito, S., & Joslyn, M. A. (1964). Presence of several phenolic components in fruit proanthocyanidins. *Nature, 204*, 475–476.

Ito, S., & Oshima, Y. (1962). Studies on the tannin of Japanese persimmon (*Diospyros kaki* L.). Part I: Isolation of leucoanthocyanin from kaki fruit. *Agricultural and Biological Chemistry, 26*, 156–161.

Iwamoto, M. (2013). Seichōzai no rekishi-teki haikei to seishu e no ōyō (Historical background of the finning agent, kakishibu and its application to sake). *Journal of the Brewing Society of Japan, 108*(5), 318–325.

Kanbayashi, S., & Kanbayashi, H. (2017). *Uji cha to Kanbayashi ichizoku (Uji tea and the Kanbayashi clan)*. Miyaobi Publishing.

Kawakami, K., Aketa, S., Nakanami, M., Iizuka, S., & Hirayama, M. (2010). Major water-soluble polyphenols, proanthocyanidins, in leaves of persimmon (*Diospyros* kaki) and their α-amylase inhibitory activity. *Bioscience, Biotechnology, and Biochemistry, 74*(7), 1380–1385.

Kennedy, J. A., & Jones, G. P. (2001). Analysis of proanthocyanidin cleavage products following acid-catalysis in the presence of excess phloroglucinol. *Journal of Agricultural and Food Chemistry, 49*, 1740–1746.

Komatsu, S., & Matsunami, N. (1924). On kakishibu I: Constitution of shibuol I. *Memoirs of the College of Science, Kyoto Imperial University, Series A, 7*, 15–23.

Komatsu, S., & Matsunami, N. (1925). On kakishibu III. *Memoirs of the College of Science, Kyoto Imperial University, Series A, 8*, 231–240.

Komatsu, S., Matsunami, N., & Ishimasa, M. (1925). On kakishibu II. *Memoirs of the College of Science, Kyoto Imperial University, Series A, 8*, 43–49.

Konno, K. (1996). Taberaremai to suru shokubutsu to tabeyo to suru konchu no kobo (The offense and defense between plants trying not to be eaten and insects trying to eat). *Kagaku To Seibutsu, 34*(9), 580–584.

Kyoto Furitsu Yamashiro Kyodo Shiryokan (Kyoto Prefectural Yamashiro Local History Museum). (1997). *Kakishibu no chikara: Kurashi to dentōkogei o sasaeta Minamiyamashiro no Kakishibu The power of Kakishibu—Supporting life and traditional crafts in Minami-Yamashiro region)*. Shobunsha Publishing.

Kyoto Furitsu Yamashiro Kyodo Shiryokan (Kyoto Prefectural Yamashiro Local History Museum). (2009). *Amagaki, hoshigaki, kakishibu: Minami-Yamashiro no kaki saibai to kakishibu bunka (Sweet persimmon, dried persimmon, and kakishibu—Persimmon cultivation and kakishibu culture in Minamiyamashiro region)*. Shobunsha Publishing.

Li, C., Leverrence, R., Trombley, J. D., Xu, S., Yang, J., Tian, Y., Reed, J. D., & Hagerman, A. E. (2010). High molecular weight persimmon (*Diospyros kaki* L.) proanthocyanidin: A highly galloylated, A-linked tannin with an unusual flavonol terminal unit, myricetin. *Journal of Agricultural and Food Chemistry, 458*, 9033–9042.

Liu, S. X., & White, E. (2012). Grape seeds proanthocyanidins: An overview of in vivo bioactivity in animal models. *Nutrients, 11*, 2435–2453.

Lv, T.-T., Zhao, O., & He, J.-R. (2024). Chemical composition and antioxidant capacity of proanthocyanidins from Chinese quince (*Chaenomeles sinensis*) fruit at different growth stages. *Journal of Food Measurement and Characterization, 18*(3), 1–13.

Matsuo, T., & Ito, S. (1977). Kakitannin nitsuite (On persimmon tannin). *Kagaku to Seibutsu (Chemistry and Biology), 15*(11), 732–736.

Matsuo, T., & Ito, S. (1978). The chemical structure of kaki-tannin from immature fruit of the persimmon (*Diospyros kaki* L.). *Agricultural and Biological Chemistry, 42*(9), 1637–1643.

Matsuo, T., & Ito, S. (1981a). Comparative studies of condensed tannins from several young fruits. *Journal of the Japan Society for Horticultural Science, 50*, 262–269.

Matsuo, T., & Ito, S. (1981b). A simple and rapid purification method of condensed tannins from several young fruits. *Agriculture and Biological Chemistry, 45*(8), 1885–1887.

Matsuo, T., Shinohara, J., & Ito, S. (1976). An improvement on removing astringency in persimmon fruits by carbon dioxide gas. *Agricultural and Biological Chemistry, 40*, 215–217.

Matsuo, T., Tamaru, K., & Ito, S. (1984). Chemical degradation of condensed tannin with phloroglucinol in acidic solvents. *Agricultural and Biological Chemistry, 48*(5), 1199–1204.

Matthews, S., Mila, I., Scalbert, A., Pollet, B., Lapierre, C., Hervé du Penhoat, C. L. M., Rolando, C., & Donnelly, D. M. X. (1997). Method for estimation of proanthocyanidins based on their acid depolymerization in the presence of nucleophile. *Journal of Agricultural and Food Chemistry, 45*, 1195–1201.

McMurrough, I., & McDowell, J. (1978). Chromatographic separation and automated analysis. *Analytical Biochemistry, 91*, 92–100.

Miyamoto, T. (2023). Shokubutsu polyphenols no kōkin kōdokuso sayō kikō ni kansuru kenkyū (Studies on mechanisms for antibacterial and antitoxin actions of plant polyphenols). *Japanese Journal of Food Microbiology, 40*(4), 103–112.

Murry, N. J., Williamson, M. P., Lilley, T. H., & Haslam, E. (1994). Study of the interaction between salivary proline-rich proteins and a polyphenol by ^{1}H-NMR spectroscopy. *European Journal of Biochemistry, 219*(3), 923–935.

Marfella R, Prattichizzo F, Sardu C, Fulgenzi G, Graciotti L, Spadoni T, D'Onofrio N, Scisciola L, La Grotta R, Frigé C, Pellegrini V, Municinò M, Siniscalchi M, Spinetti F, Vigliotti G, Vecchione C, Carrizzo A, Accarino G, Squillante A, Spaziano G, Mirra D, Esposito R, Altieri S, Falco G, Fenti A, Galoppo S, Canzano S, Sasso FC, Matacchione G, Olivieri F, Ferraraccio F, Panarese I, Paolisso P, Barbato E, Lubritto C, Balestrieri ML, Mauro C, Caballero AE, Rajagopalan S, Ceriello A, D'Agostino B, Iovino P, Paolisso G. (2024) Microplastics and Nanoplastics in Atheromas and Cardiovascular Events. *New England Journal of Medicine, 390(10), 900-910.*

Ngure, F. M., Wanyoko, J. K., Mahungu, S. M., & Shitandi, A. A. (2009). Catechins depletion patterns in relation to theaflavin and thearubingins formation. *Food Chemistry, 115(1), 8–14.*

Nakabayashi, T. (1968). Kakishibu no kagaku (Chemistry of kakishibu). *Journal of the Society of Brewing, Japan, 63*(11), 1149–1154.

Nakabayashi, T. (1971). Kakishibu no dekirumade (How kakishibu is made). *Journal of the Brewing Society of Japan, 66*, 474–476.

Nakamura, K., Ishiyama, K., Sheng, H., Ikai, H., Kanno, T., & Niwano, Y. (2015). Bactericidal activity and mechanism of photoirradiated polyphenols against gram-positive and -negative bacteria. *Journal of Agricultural and Food Chemistry, 63*, 7707–7713.

Neilson, A. P., O'Keefe, S. F., & Bolling, B. W. (2016). High-molecular-weight proanthocyanidins in foods: Overcoming analytical challenges in pursuit of novel dietary bioactive components. *Annual Review of Food Science and Technology, 7*, 43–64.

Nunokawa, Y., & Mena, E. (1973). Seishu seityo-yō Kakishibu ni kansuru kenkyū (Study on persimmon tannin for sake filtration (Report 5) estimation of particle size of persimmon tannin and sake protein turbidity). *Journal of the Brewing Society of Japan, 68*(9), 708–711.

Ohnishi-Kameyama, M., Yanagida, A., Kanda, T., & Nagata, T. (1997). Identification of catechin oligomers from apple (*Malus pumila* cv. Fuji) in matrix-assisted laser desorption/ionization time-of-flight mass spectrometry and fast-atom bombardment mass spectrometry. *Rapid Communications in Mass Spectrometry, 11*, 34–36.

Oszmianski, J., & Bourzeix, M. (1996). Comparison of methods for determining the content and polymerization of proanthocyanidins and catechins. *Polish Journal Of Food And Nutrition Sciences, 5*, 43–50.

Parker, L., Daly, N., & Royte, E. (2018). Umi ni nagarederu plastics (Plastics flowing into the sea). *National Geographic Japan Edition*, (June issue), 32–69.

Piretti, M., Pistore, R., & Razzoboni, C. (1985). The chemical constitution of kaki (*Diospyros kaki*) tannin. *Annali di Chimica, 75*, 137–144.

Rodríguez-Pérez C, García-Villanova B, Guerra-Hernández E, Verardo V. (2019). Grape Seeds Proanthocyanidins: An Overview of In Vivo Bioactivity in Animal Models. *Nutrients, 11*, 2435–2453.

Saito, K. (2018). Cha catechins to sono yūkō riyō (Characteristics and effective use of tea catechins). *Agriculture and Horticulture, 93(2), 108–115.*

Sakaguchi, T., Ueda, K., & Kawabata, R. (2014). Kakishibu de virus taisaku kakitannin niyoru kyōryoku na virus fukatsuka sayō (Virus countermeasures with Kakishibu: Strong virus inactivation effect by persimmon tannin). *New Food Industry, 56*, 17–24.

Scalbert, A., & Haslam, E. (1987). Chemical defense and polyphenols of Quercus robur. *Phytochemistry, 26*, 3191–3195.

Schofield, P., Mbungua, D. M., & Pell, A. N. (2001). Analysis of condensed tannins: A review. *Animal Feed Science and Technology, 91*, 21–40.

Shi, L., Wang, J., & Lv, Y. (2021). Research progress on analysis method of procyanidins. *Earth and Environmental Science, 705*, 1–10.

Shi, L., Lv, Y.-K., Wang, J.-W., Yang, L.-L., & Shen, S.-G. (2022). A novel thiolysis-high-performance liquid chromatography method for the determination of proanthocyanidins in grape seeds. *Journal of Separation Science, 45*(11), 1874–1883.

Souquet, Jean-Marc & Cheynier, Veronique & Brossaud, Franck & Moutounet, Michel. (1996). Polymeric proanthocyanidins from grape skins. *Phytochemistry, 43*(2), 509-512.

Spencer, K. C. (1988). *Chemical mediation of coevolution. Academic Press.*

Stringano, E., Gea, A., Salminen, J.-P., & Mueller-Harvey, I. (2011). Simple solution for a complex problem: Proanthocyanidins, galloyl glucoses and ellagitannins fit on a single calibration curve in high performance-gel permeation chromatography. *Journal of Chromatography A., 1218*, 7804.

Takemori, K., Akaho, K., Iwase, M., Okano, M., & Kometani, T. (2016). Effects of persimmon fruit polyphenols on postprandial plasma glucose elevation in rats and humans. *Journal of Nutritional Science and Vitaminology, 68*, 331–341.

Takemori, K., Okano, M., Shimeno, A., & Kometani, T. (2022). Polyphenol composition analysis in 30 cultivars of persimmon fruits; Comparison of polyphenols between cultivars in Japan, China, and Korea. *Food Science and Technology Research, 28*(5), 403–413.

Tanaka, T., & Koumoto, I. (2003). Oxidation of tea catechins: Chemical structures and reaction mechanism. *Food Science and Technology Research, 9*, 128–133.

Tanaka, T., Takahashi, R., Kouno, I., & Nonaka, G. (1994). Chemical evidence for the de-astringency (insolubilization of tannin) of persimmon fruit. *Journal of the Chemical Society, Perkin Transactions, 1*, 3013–3022.

Thompson, R. S., Jacque, D., Haslam, E., & Tanner, J. N. (1972). Plant proanthocyanidins. Part I: Introduction; the isolation, structure, and distribution in nature of plant procyanidins. *Journal of the Chemical Society, Perkin Transactions, 1*, 1387–1399.

Ueda, K., Kawabata, R., Irie, T., Nakai, Y., Tohya, Y., & Sakaguchi, T. (2013). Inactivation of pathogenic viruses by plant-derived tannins: Strong effects of extracts from persimmon (*Diospyros kaki*) on a broad range of viruses. *PLoS One, 8*(1), e55343.

Wang, S., Smyth, H. E., Mantilla, S. M. O., Stokes, J. R., & Smith, P. A. (2024). Astringency and its sub-qualities: A review of astringency mechanisms and methods for measuring saliva lubrication. *Chemical Senses, 1*, 49.

Weinges, K. (1975). *Topics in flavonoid chemistry and biochemistry*. Elsevier Science.

Weinges, K., Bahr, W., Ebert, W., Goritz, K., & Marx, H.-D. (1969). Constitution, origin, and significance of flavonoid tannins. *Progress in Chemistry of Organic Natural Products, 27*, 158–260.

Worm, B., Lotze, H. K., Jubinville, I., Wilcox, C., & Jambeck, J. (2017). Plastic as a persistent marine pollutant. *Annual Review of Environment and Resources, 42*, 1–26.

Wu, P.-W., & Hwang, L. S. (2002). Determination of soluble persimmon tannin by high performance gel permeation chromatography. *Food Research International, 35*, 793–800.

Yamane, H., Konno, K., Takabayashi, T., Sassa, T., & Oikawa, H. (2010). Chemical defense and toxins of plants. In *Comprehensive natural products II: Chemistry and biology* (Vol. 4, pp. 339–385). Elsevier.

Yamauchi, K., Soyano, M., Kobayashi, M., Kamatari, Y. O., & Mitsunaga, T. (2023). Protein aggregation model to explain the bioactivity of condensed tannins. *Food Chemistry, 441*, 135870.

Yanagida, A., Shoji, T., & Kanda, T. (2002). Characterization of polymerized polyphenols by size-exclusion HPLC. *Bioscience, Biotechnology, and Biochemistry, 66*(9), 1972–1975.

Yokoi, T., Suzuki, K., Tsuboi, K., Oyabu, Y., & Nagoya Art and Design Technical College Kakishibu Research Group. (1995). Kakishibu no kenkyū sono rekishi to dōgu *(A study of Kakishibu: Its history and tools)*. Nagoya University of Arts and Design Technical College.

Yoshida, K., Inahashi, M., Noro, F., & Murakami, H. (1985). Kaki kara bunri shita kōbo no tokusei to ishū no nai kakishibu seizō shiken (Characteristics of yeast isolated from persimmons and trial production of odorless kakishibu). *Journal of the Brewing Society of Japan, 80*(7), 471–475.

Zhang, T., Li, G., Mo, H., & Zhi, C. (2011). Persimmon tannin composition and function. *Advances in Biomedical Engineering, 1–2*, 389–392.

Zunjarrao, S. S., Tellis, M. B., Joshi, S. N., & Joshi, R. S. (2020). Plant–insect interaction: The saga of molecular coevolution. In *Co-evolution of secondary metabolites* (pp. 19–45). Springer.

Tomoaki Matsuo is an emeritus professor at Kagoshima University in Kagoshima City, Japan. He earned his Doctor of Agriculture degree from Kyushu University's Faculty of Agriculture for his research achievements, which included analyzing the chemical structures of astringent components in astringent persimmons and condensed tannins in other fruits. Following his retirement, he established the Scientific Research Association for Kakishibu and Kakitannin. He routinely conducts Kakishibu lectures and exhibitions, organizes workshops, maintains a website, and publishes the Kakishibu Newsletters. Through these activities, he disseminates information regarding "the wonders and appeal of Kakishibu, and its industrial applications" both domestically and internationally. Concurrently, he is involved in technological development aimed at new applications of Kakishibu.

Part III
Business and Industry of Kakishibu

Chapter 5
Opening New Frontiers in the Kakishibu Industry Through Green Innovation

Tomoaki Matsuo

Abstract In this chapter, we first address the major consumption of Kakishibu throughout Japanese history, and after these historical swells, Kakishibu industry is in decline at a present time, even with its high industrial potential. Pointing out and discussing the current issues facing the Kakishibu industry are as follows: production expansion of raw astringent persimmon production, harvesting of green immature persimmon fruit in midsummer, gelation of persimmon juice in the course of fermentation and long-aging, generation of strong acetic acid odor and putrid odor. Subsequently, drawing on the development of the global tannin industry, we discuss the possibilities for developing new applications for high molecular weight (HMW) tannins including Kakishibu.

From a different perspective on the development of the Kakishibu industry, this report details the establishment of the Science Research Association for Kakishibu and Kakitannin and the various regional activities it has undertaken, which have facilitated the sharing of information from many different viewpoints. Next, referring to traditional knowledge related to Kakishibu, we explore creating new natural materials that meet the needs of modern society without relying on radical chemical reactions. To evoke an image of building a slow-life society utilizing natural polymer compounds, we illustrate and introduce the concept of "Kakishibu Village." Finally, we explain the significance of a slow-life society centered on new materials created through the green transformation of Kakishibu.

Keywords Green transformation · Kakishibu industry · Kakishibu Village · Shibugami · Shibuzumi painting · Tannin industry · The Scientific Research Association for Kakishibu and Kakitannin · Washi

T. Matsuo (✉)
The Scientific Research Association for Kakishibu and Kakitannin, Kyoto, Japan
e-mail: kaki@tannin.info

K. Kano et al. (eds.), *Traditional Knowledge for Green Innovation with Japanese Persimmon Tannin, Kakishibu*, https://doi.org/10.1007/978-981-95-7805-4_5

5.1 Issues Regarding the Current State and Future of the Kakishibu Industry

Kakishibu, made by squeezing the juice from immature astringent persimmons, has supported various aspects of traditional Japanese life. From the Edo period (1603;2013;1868) to around 1955, many common people 2019;s gardens had astringent persimmons, and they were at the center of daily lives. Looking back at the periods when Kakishibu was consumed in large quantities, the first phase was likely when it was used aroud traditional fishing grouds for fishing lines, nets, dragnets, or work clothes on boats. Although there are no precise records in the literature, this mass consumption continued for a long time, from the Middle Ages in Japan, or even earlier, until the mid-Showa period. The second phase is thought to have started in the Sengok Period in Japanese history. It is believed that early mountain castles and checkpoints in Japan began to be built around the middle of the Muromachi period, during the Sengoku period, but it is not clear whether Kakishibu (or shibuzumi painting) were used for their wooden walls and fences. No clear documents have been found. From the early to mid-Edo period, the wooden walls of many castles, checkpoints, and private homes were popularly painted black to enhance water resistance and durability and prevent decay. These were treated with a paint made by mixing Kakishibu and soot from burning pine wood, known as *shibuzumi* painting. The amount of Kakishibu used for this paint throughout Japan is expected to have been considerable.

The third phase took place in the mid-Showa period. At this time, Japan experienced high economic growth, and the consumption of alcohol, especially sake (Japanese rice wine), was greatly increasing. As a result, Kakishibu was consumed in large quantities for *orisage,* dreg- and turbidness-lowering treatment in the production of sake to prevent cloudiness after filtering and bottling. Additionally, it was used to treat various tools for transporting each ingredient in the sake-making process and *sakabukuro*, cloth bags used at pressing stage of making sake. At this time, the consumption of Kakishibu was so high that it reportedly caused a major shortage of domestically produced astringent persimmons.

This third phase has lasted for a long time despite gradually declining the consumption, and then at present time the fourth phase may come with the setting of the Sustainable Development Goals (SDGs) and many people around the world turning their attention to environmental issues, there is increasing focus on natural and organic materials instead of plastic, and we are hoping for the growing interest in Kakishibu as a pioneering natural material.

The condensed tannin contained in Kakishibu is a unique natural material with various characteristics and is a high molecular weight (HMW) substance that is friendly to humans and the environment. Given these useful properties, it is necessary to pay renewed attention to this unique natural HMW tannin, and reflect on the traditional Kakishibu culture, in order to actively utilize it again. However, the Kakishibu industry is now facing a dilemma between scaling up of demand and supply. Activities and businesses that create large demand for Kakishibu often stop because their supply capacity is questioned. On the other hand, efforts to supply in larger scale often require outlets for sales and consumption in advance. Naturally, producers of astringent persimmons cannot increase production of something that is not guaranteed to sell. As a result, these two challenges are like two flat wheels of a car that now cannot move. In order to enable the car to start moving again, both wheels need to move in balance with harmony. In other words, developing a new Kakishibu industry requires a reliable supply of, and high demand for, high-purity, high-quality Kakishibu.

To consider its future growth, we need to ensure the efficient management of Kakishibu's entry and exit into markets: the stable production of astringent persimmons as raw materials, and the consistent manufacturing and supply of Kakishibu, including squeezing, fermentation, aging, storage, sales, transportation, and both small-scale and large-scale use. Additionally, it is important to identify and develop markets where Kakishibu is used and consumed in large quantities. In Japan, given the declining birth rate and aging population, we need to design a circular economy centered on Kakishibu that incorporates labor-saving measures. In the remainder of this section, the current issues at each stage of the Kakishibu-related industry will be discussed in more detail.

5.1.1 Expanding Production of Astringent Persimmons for Raw Material of Kakishibu

Almost all of the persimmons produced in the world, including dried persimmons, are for eating, and the countries with the highest production volumes are China, Korea, Azerbaijan, Spain, and Japan, in descending order (Japan Fruit Association, 2022). However, the production of astringent persimmons for processing other than food is hardly captured in current statistical data. In Japan, recent annual data on the total production of persimmons over a 5-year period (FY2019-FY2023) averaged 198,050 tons, and the roughly same amount of sweet persimmons was produced as astringent persimmons (Ministry of Agriculture, Forestry and Fisheries, 2019, 2020, 2021, 2022, 2023). With respect to consumption, for astringent persimmons this includes fresh consumption such as de-astringent persimmons (persimmons from which the astringency has been artificially removed) and dried persimmons. There are no precise official statistics on the annual production of Kakishibu in Japan, but a rough estimate, taking into account *shibuya*, traditional small-scale Kakishibu manufacturing and sales companies, along with local small groups and individual producers, suggests that about 180 tones or less of raw persimmons are harvested and juiced, yielding roughly 90 tons of Kakishibu.

Considering the current annual production of astringent persimmons, the use of abandoned farmland, declining demand for astringent persimmons and labor shortages, there are several benefits to growing raw persimmons for Kakishibu compared to astringent persimmons for eating, and the total production may rise in the future.

Compared to growing persimmons for eating, the annual work hours for growing raw persimmons for Kakishibu are much lower because this work involves almost no tasks like early bud removal, fruit thinning, pollination, pesticide spraying, bagging before harvest, production adjustments, and packaging after harvest. The environmental friendliness is a notable benefit. Also, it is a big advantage that it does not matter if the fruit is slightly damaged during growth or at harvest, or if the fruit is deformed. Given these benefits, it is not difficult to increase the production of raw persimmons for Kakishibu, which is quite easy to cultivate and harvest, and it is thought that a production volume of 500 tons per year can be easily achieved in the short term. Approximately half the weight of fresh persimmons is juice, and after fermentation and long aging, most of this juice becomes Kakishibu. Since this contains about 5% condensed tannin (based on multiple interviews with *shibuya* conducted by the author and refer to Sects. 4.3.2.1 and 4.3.3 in the Chap. 4), it is estimated that approximately 12,500 kg of dried tannin powder can be obtained annually. With this amount, if you were to make *shibugami* (Kakishibu-treated *washi*), for example, it has been reported that if you apply Kakishibu by hand with a brush on handmade *washi*, about 2.0~3.0 g of Kakishibu will adhere per square meter (Oda, 2014; Sakate & Kojima, 2015). Using these numbers as estimates, and considering that it is usually applied at least twice to avoid uneven coating (namely, 4.0~6.0 g), in order to make a roll of 2 m wide and 1000 m long, one would need 8 kg to 12 kg of Kakishibu. This amount will vary greatly depending on the type of *washi* and how it is made (whether handmade or machine-made). It is also thought to vary greatly depending on the type and concentration of Kakishibu and its viscosity. Of course, an important initial question is what kind of *shibugami* to make.

There are generally two types of astringent persimmons used to make Kakishibu. The first is the relatively large astringent persimmon, represented by varieties such as "*Atago*" and "*Saijo*," which are widely grown in western Japan for fresh eating, and they are also harvested early in their growth when still small and immature (higher in tannin content) and used as a raw material for Kakishibu. The second type of astringent persimmon, represented by varieties such as "*Tsurunoko*," "*Tenno*," "*Horenbo*," and "*Mame*," are relatively small and are harvested solely for use as a raw material for Kakishibu. In some regions, the small, ripe, and colored fruits are sometimes processed into dried persimmons.

Most of these varieties have not been systematically produced for Kakishibu production on a large scale yet. Many are naturally grown trees, and may grow up to 5~10 m. This requires a lot of manpower for harvesting, and working at such heights can be difficult and dangerous. It is also very hard work to pick a lot of small immature fruits during the peak of summer heat (Fig. 5.1); this is currently the most technically demanding task. It is not an exaggeration to say that picking during summer is one of the significant challenges of the Kakishibu industry.

An illustration of the autumn harvest time is shown in Fig. 5.1 (left), which describes the dangerous work at a higher place. Regarding the harvest of raw persimmons for Kakishibu, the same dangerous work should be done in very hot midsummer because green and small persimmons are required for producing Kakishibu. In some places in Japan, small astringent persimmons are harvested by breaking off

Fig. 5.1 Harvesting of fruits at high places of grown persimmon trees. Harvest process in autumn (left); the raw material for Kakishibu, "*Tenno*" persimmon in midsummer (right)

whole twigs using special tools, as shown in the illustrations (left). This illustration was reprinted from a booklet provided with permission by the Kyoto Prefectural Yamashiro Local History Museum, Kyoto.

I would like to propose that, just as grapes are divided into those for fresh consumption and those for wine production, each being produced independently, the raw persimmons for Kakishibu could also be processed as astringent persimmons, with their own varieties and cultivation methods, and cultivated and harvested as completely separate entities. In particular, it would be necessary to design the orchard, weed control, irrigation, and harvest in advance with a focus on labor-saving measures. For example, it is essential to consider the width for introducing machinery between trees by keeping the tree height to about two or less meters.

Additionally, how to pick each fruit from the branch is a challenge involving how to harvest large quantities of fruit quickly using labor-saving technology and bring them to warehouses. Unlike persimmons for eating, a major advantage of raw persimmons for Kakishibu is that they can be harvested in bulk even if slightly damaged or uneven in shape. When planting new persimmon trees, using 2- or 3-year-old grafted seedlings, one can expect they will start producing fruit in about 5 years, with a sufficient harvest in 10 years. Depending on how well the trees are cared for, harvests can occur consistently every year for about 40 to 50 years. If we can aim to make the total annual production in Japan 10,000 tons as a first step, we may be able to turn it into a growing industry.

Furthermore, securing manpower to pick the fruit is a significant issue. Creating a system like the owner system of raw persimmon trees, where people contracted with the farmer to have a right to harvest the persimmon trees, or fun volunteer activities such as agricultural experiences, tourist farms, is also a powerful attempt to involve young people in big cities.

In astringent persimmon orchards facing current conditions; many overgrown trees, one short-term solution to maintain a certain yield while avoiding excessive labor and hazardous work involves adjusting their trees height to a manageable level by cutting the main trunk.

Various other ideas can be considered to expand the production of astringent persimmons for the raw material of Kakishibu. For example, the introduction of greenhouse and hydroponic cultivation, robot harvesters, or fruit thinning technology can be expected in the future. The fruit thinning technology during the early growth stage of sweet persimmons is already being tested at several sites, and there may be treatment conditions suitable for harvesting raw persimmons for Kakishibu (Yamamura & Naito, 1975; Naito et al., 1975).

Similarities and Differences Between Red Wine Brewing and Kakishibu Production

Kakishibu is similar to red wine and is a fermentation product of persimmon juice. The fruits used as raw materials for both products are cultivated and harvested separately from fruits for raw consumption. Some of the astringent persimmons used as raw materials after full maturation and coloration may be used for raw consumption. For Kakishibu, the harvest time is completely different. That is, immature fruits that have not yet ripened are harvested in midsummer. Traditionally, after harvesting, the sugar in both fruit juices is naturally fermented by the yeast present on the surface of the fruit, producing foam and carbon dioxide, and then long-aging and maturing proceeds. The fermentation and aging of both are quite different, and the chemical components contained are different. Both may also undergo a sterilization process called pasteurization. Wine mainly contains ethanol, non-volatile organic acids, and tannins, but Kakishibu, especially old and aged one, contains almost no ethanol and is higher in tannins and volatile organic acid, mainly acetic acid. In the case of wine, most of the astringency and red pigments come from the fruit skin and seeds, but in the case of Kakishibu, the astringency and dark black-brown color come from the tannins accumulated in the fruit. The end product purposes of both are completely different. Wine is naturally a beverage and color, aroma, and taste are important with products carefully selected and made, while Kakishibu is used as dye, paint, a precipitant of proteins, deodorant, and discoloration preventer. It also has a variety of characteristics as polyphenols and tannins and is used in various fields. Finally, the ways both are used are generally completely different.

5.1.2 Harvesting, Accumulation, and Delivery of Raw Persimmons

Most of the raw persimmons for Kakishibu are either young fruits or immature fruits. As such, they are green and growing, and the fruits are easily damaged. If the temperature is high after harvest, they will wilt easily, soften and become flabby, and washing and squeezing will not go well. As much as possible, it is desirable to plan and pick early in the morning, keeping them cool, and then collect them for transport to the factory. It is not desirable to pile up the fruit and pack it in a container for a long time. It is also desirable to avoid vibration as much as possible. Currently, bags with a coarse mesh, as opposed to box-shaped containers, are sometimes used. In the Kakishibu manufacturing plant, it is important to wash with cold water, to quickly lower the product temperature, and squeeze. It is necessary to avoid using iron machines, tools, and containers for work, as iron ions turn the Kakishibu liquid black. Therefore, stainless steel is preferred. Naturally, it is necessary to remove iron ions from the water used within the factory as well.

5.1.3 Squeezing, Fermentation, Aging, Storage and Delivery of Produced Kakishibu

There are two major challenges in the traditional production and storage of Kakishibu. The first challenge is the gelation of the squeezed juice. Based on experimental observation of Kakishibu and the experience of *shibuya*, traditional small-scale Kakishibu manufacturing and sales companies, Kakishibu under acidic conditions and with high concentrations (4% tannin content or more) that is left in a cool, dark place for a long time, starts to become muddy and begins to solidify into a gel, eventually becoming a agar-like gel (gelation occurs). The harvested and collected persimmons are washed, and they are squeezed and the resulting juice is fermented with yeast. When the fermentation bubbles stop, the fermentation step moves to aging in barrels or tanks. In this process, gelation often occurs by chance, and the solution in the barrel or tank may gel and solidify. As an example, we have heard that at a *shibuya*'s factory Kakishibu occasionally solidified inside 6,000-liter aging/storage tanks, necessitating disposal and resulting in significant losses. Moreover, it has been observed that there are considerable differences in behavior between *shinshibu,* new/ fresh Kakishibu, and *koshibu,* old/aged Kakishibu.

Currently, the following measures are considered against gelation:

1. Lower the concentration of condensed tannin in Kakishibu (make it less than 4%).
2. Store at high temperatures (35–60°C or more). However, in this condition, the color of the Kakishibu gradually darkens with higher temperatures and becomes a dark black-brown.
3. Stir continuously or apply vibration.
4. Raise the pH, to 6.5-8.5. Kakishibu is prone to gelation when acidic. However, on the alkaline side, the color becomes a dark black-brown, and oxidation and decomposition are likely to occur.
5. Decompose the molecules of condensed tannin in Kakishibu and transform them low molecular weight (LMW) compounds, corresponding to the purpose of use.
6. Add LMW compounds that inhibit hydrogen bonding and hydrophobic bonding between condensed tannin molecules in Kakishibu.

Kakishibu that has gelled can generally be dissolved by heating to 85 °C or higher and stirring or by stirring in an alkaline state, but various side reactions, including decomposition, will occur as the color gradually darkens to a dark black-brown.

The second major challenge is that, due to long-term fermentation and aging, Kakishibu develops a strong acidic and rotten smell. As a result, there are sometimes pollution issues around the Kakishibu factory, and workers have to cope with the terrible odor all year round. So far, the following measures have been taken to reduce the strong odor:

1. Use new yeast that is less likely to produce a bad smell after heat sterilization of the juice.
2. Production of Kakishibu (*shinshibu* and *kishibu*) that has a short aging period after alcohol fermentation (the color is light yellow).
3. Deodorize the Kakishibu produced by traditional methods through (1) using dialysis methods, treatment with ultrafiltration membranes (Iwamoto, 2013) or ion exchange resins, and (2) dry the Kakishibu to be powder or granule by dry-spraying or freeze-dryng.
4. Nara-style Kakishibu, which is different from traditional Kakishibu and known as having no fermentation process involved, has a unique, less pungent smell (Hamazaki, 2010).

5.1.4 Stability, Natural Degradation, and Decomposition of Kakishibu

Kakishibu gradually changes color over the fermentation and aging periods, or when stored for a long period of time, depending on the conditions. In particular, when Kakishibu is neutral to slightly alkaline and exposed to air, oxidation occurs and the color progresses to a dark black-brown. When Kakishibu is applied to *washi* or wood and dried, the color of the Kakishibu becomes darker. If it is exposed to direct sunlight for a long period of time, it decomposes and becomes whitish. Essentially, it is difficult to use Kakishibu on its own outdoors. The main component of Kakishibu, condensed tannin, absorbs ultraviolet rays due to its structure, gradually deteriorates, and decomposition progresses. Since fading progresses under direct sunlight, regular repainting is required on walls and roofs.

Another method to stabilize Kakishibu is to mix Kakishibu with fine mineral powder that reflects without absorbing direct sunlight, and it has been found to last longer when applied. For example, in the past, techniques such as mixing in straw or wood ash, or incorporating minerals (metal salts) like *bengara,* red iron oxide, were used. It is an interesting problem to work out how quickly Kakishibu-treated paper (*shibugami*) and Kakishibu products decompose in different environments.

5.2 The World's Tannin Industry and Japanese Kakishibu Industry

Historically, the production of tannins from trees began as part of the urgent process to create stable leather products. In medieval Europe, to keep tannin for tanning, they cut down forest trees and used hot water to extract the bark, heartwood scraps, and chips. Nevertheless, the demand for leather shoes, horse gear, and other military items increased due to wars, and the industry that tanned rawhide grew significantly. Many forest trees were felled due to the rapid development of this tannin industry. When the forest timber was eventually used up, Western tannin merchants turned their eyes to tropical rainforest trees and timber, and secured new tannin raw materials by cutting down these trees. Many trees and timber growing in tropical and subtropical regions have a high tannin content and can provide tannin more efficiently. Therefore, large-scale deforestation for the extraction of tannin has accelerated in these regions. Most of this logging was accompanied by the manufacture of building materials, securing agricultural land, and excessive fuel collection.

The use of Japanese Kakishibu, water-soluble tannin, is a very environmentally friendly and eco-friendly material as a tannin raw material. Since the raw material is green, immature fruit, it is environmentally and socially beneficial because it can be harvested reliably each year while keeping the trees intact. Also, the average tree lifespan is well over 50 years, so a certain amount of production can be expected for at least 40 years. For existing cultivation methods, the average production per tree is around 40 kg, but can vary greatly depending on the variety of astringent persimmon, land, cultivation, and harvesting method. With about 60 planting trees per 10 a (acre), which is 1000 m^2, the average expected harvest is about 2.4 tons. If half of the weight is squeezed into juice, then it leads to about 1.2 tons, and if 5% of that is Kakishibu, it is roughly calculated that 60 kg of tannin dry powder can be obtained from 10 acre of land every year (Fujishima et al., 2011; Seino et al., 2016).

To provide a sense of comparison, let us look at the industrial use of HMW tannins extracted and prepared from other raw materials in the world, especially bark, heartwood, and other natural plant materials, in discussing the future of Kakishibu (kakitannin) application development. Currently, the global market size for tannins is expected to reach $4 billion (600 billion yen) in 2030, growing at a Compound Average Growth Rate (CAGR) of 5.7% from 2021 to 2030 (Grand View Research, Tannin Market Size, Share & Trends Analysis Report 2020). The growth of the global tannin market is mainly supported by demand from the leather tanning, wine, paint and coating, and adhesive manufacturing industries. In particular, it is widely used in applications such as luxury leather interiors for automobiles, clothing, shoes, healthcare, wood adhesives, and anti-rust primers (Grand View Research, Tannin Market Size, Share & Trends Analysis Report 2020). Recently, its use as a food additive (for antioxidant and antibacterial effects) has been attracting attention. Therefore, globally, tannins can be considered a growth industry, and for a wide range of industrial uses of tannins in the West new applications are being developed. Prizzi (2019) published an excellent review detailing the origin and historical development of tannins, as well as their long-standing and future important uses. In the review, he argued that the diverse uses included not only leather tanning agents and wood adhesives, but also additives for wine, beer, and juice, fire-resistant and insulating porous resins, coagulation and removal agents for heavy metals, metal corrosion inhibitors, and raw materials for pharmaceuticals and medical use (Pizzi, 2019). What is also remarkable, Singh and Kumar also reviewed on the application of tannins derived from various plant materials in modern industry and their utilization as a sustainable raw material in emerging industries (Singh & Kumar, 2020). In addition, Neto et al. (2020) have discussed in detail the role of tannin (here, HMW proanthocyanidin) in the formation of a sustainable society, especially for reducing environmental impact. If the cost of Kakishibu can be justified, it is thought that it could be used in sustainable society. What is interesting is that in Japan, Kakishibu or kakitannin has often been used in its natural state in daily life and industry. Even when Kakishibu was advanced, it was only mixed with other natural materials, without strong chemical modification. This is a relative advantage over more chemically intensive approaches. For example, making radical chemical modifications to develop new applications is undesirable because problems may arise with decomposition after use. Also, it is a problem if excessive energy or highly reactive chemicals are required at the manufacturing stage (Neto et al., 2020). Therefore, in the future, we should proceed with tannin development while considering whether all the processes from the production, use, and disposal of tannins are sufficiently human and environmentally friendly and in line with the SDGs.

5.3 The Establishment and Regional Activities of the Scientific Research Association for Kakishibu and Kakitannin

In historical periods without plastic, it was Kakishibu that continued to support traditional Japanese life in all aspects, and a sustainable society centered on it was formed. Therefore, we believe that it is extremely important to pass on to the current youth and future generations the history and culture of Kakishibu, including its appeal as a natural material, and its unique characteristics as a natural HMW compound. I think that it is considered to have world cultural heritage value that should be passed on to future generations. Preserving existing Kakishibu products, passing down their production and usage methods, and promoting awareness to the general public are urgently needed. This involves conveying and preserving each process associated with the cultivation of persimmon, the raw material of Kakishibu, to the use of Kakishibu backed by scientific technology.

In the previous section, we discussed in detail the challenges of the Kakishibu industry. In addition to that, it is important to increase social awareness of the word Kakishibu. However, with the decline of the Kakishibu industry, its wider familiarity has drastically decreased. For example, there had been keywords such as

Kakishibu or *shibuya*, traditional small-scale Kakishibu manufacturing and sales companies, in the occupational telephone directory in Japan. However, with the recent decline of Kakishibu-related businesses, these keywords disappeared at some point, and even the telephone directory has disappeared from society. Therefore, it is necessary to increase social recognition in order to pass on the valuable history, culture, and industry of Kakishibu into the future. One important part of this is creating a novel place for exchange of opinions and technology among craftsmen, technicians, and researchers currently engaged in the Kakishibu industry.

I moved to Kyoto Prefecture after retiring from the Graduate School of Agriculture at Kagoshima University in March 2013. Shortly after, I had the opportunity to meet with Professor Hino Motosugi of the Department of Fruit Science at Kyoto Prefectural University, and we discussed Kakishibu. Upon hearing that Professor Motosugi had been conducting Kakishibu hands-on workshops for the general public, we agreed to create a research group on Kakishibu. In this region, the southern part of Kyoto Prefecture, the production of Kakishibu has traditionally continued, and traditional industries that use them are also thriving. In addition, the surrounding area is a land with a high entrepreneurial spirit for new industries in the Keihanna Science Research Area.

We first gathered volunteers with various occupations in the area and created a secretariat for the research group on Kakishibu. Then, we established the Scientific Research Association for Kakishibu and Kakitannin on May 14, 2014. This association was intended to form the core of a human and informational network that enabled the inheritance and preservation of traditional culture and technology of Kakishibu, the education of young people, the nurturing and revitalization of the current Kakishibu industry, and the development of new products and new uses in the future.

As a first activity, we planned to hold a public lecture on Kakishibu about once a year to hear directly from researchers, technicians, and entrepreneurs who are currently dealing with Kakishibu and to hold an exhibition of the Kakishibu products and goods. In addition, following the recommendations from the members of the secretariat, we quickly set up a web page in January 2015 thanks to the efforts of Mr. Toshiharu Nuno. It was essential to publicize the activities of the research group to many people.

The first public lecture and Kakishibu exhibition were held at Kyotanabe City Chamber of Commerce and Industry on March 10, 2015. Despite the snowy weather, the conference room was full, with over 70 participants from not only the local area but also Fukushima Prefecture, Hyogo Prefecture, Nara Prefecture, and other places. It was very successful. After the lectures by the four speakers, we set up a time for general discussion and exchange of opinions, and created opportunities for sufficient interaction among the participants.

The third lecture we held took place on October 27, 2016, at the National Diet Library Kansai Branch's large conference room and atrium. It was supported by the Heisei 28 (2016) fiscal year regional power regeneration project grant from Kyoto Prefecture and sponsored by Seika Town, Kyoto Prefecture. We hosted a public lecture and Kakishibu exhibition focused on the production and use of Kakishibu. A

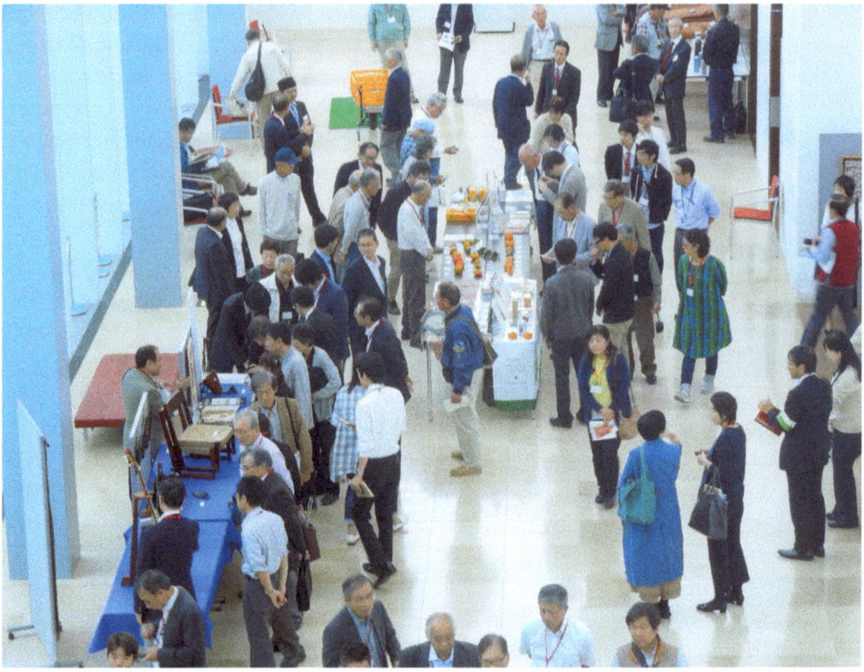

Fig. 5.2 Exhibition of a Wide Variety of Kakishibu Products held by the Scientific Research Association for Kakishibu and Kakitannin in the Atrium of the National Diet Library kansai-kan in 2018

total of 118 participants from various backgrounds and professions attended, not only from the local area (Fig. 5.2). Further, on July 8, 2018, we held the fifth lecture and exhibition at the Kyoto Hall, the alumni hall of Ryukoku University in front of Kyoto Station. The general moderator of the public lecture was Professor Yasushi Nakamura of Kyoto Prefectural University. The opening remarks were given by Professor Keizo Yonemori of the Faculty of Agriculture at Ryukoku University, who co-hosted the event. Next, we received participation and greetings from Akinori Yamashita, the Deputy Governor of Kyoto Prefecture. We received warm words of encouragement regarding interest in Kakishibu and the activities of the research group. Three people participated from Kyoto Prefecture; Director of the Commerce, Labor and Tourism Department. We also received support from Kizugawa City and Seika Town, and the chairpersons of the city council, the mayor, the chairperson of the town council, and council members. In addition, the president and managing director of the Kyoto Yamashiro Agricultural Cooperative, who had been cooperating in increasing the production of raw persimmon seedlings, also participated. In total, there were 126 participants in the lecture and 18 exhibitors at the exhibition. Despite adverse conditions such as heavy rain the day before, the event was a great success, which deeply moved the entire study group's office team. Thereafter, we posted videos of Kakishibu science experiments on the study group's website, and in March 2016, we created a page with English explanations.

We eventually decided to publish and distribute "Kakishibu News" about twice a year to facilitate communication among members of the study group and to make the activities of the study group more widely known. During this period, in addition to activities supported through subsidies from local governments and universities, we transitioned to activities based on membership fees collected from general members and supportive corporate members from society as a whole. By around 2021, we had 70 general members and 20 corporate members, and received continuous support. Furthermore, to revitalize the Minamiyamashiro region, which produces a lot of raw persimmons and used to have several *shibuya*s, we named the region "Kakishibu Village" and surveyed existing *shibuya*, Kyoto Prefectural Yamashiro Local Museum, research institutions such as universities, and shops that manufacture and sell products using Kakishibu.

In another initiative, the association collected and organized a large amount of information on Kakishibu from this region, and we were able to develop a unique map (Kakishibu Map, Fig. 5.3) that depicts experiences and things related to Kakishibu in the Yamashiro region, published on February 16, 2020. This Kakishibu map is designed to make it easy for the general public and tourists who are not familiar with Kakishibu to see what they can see, buy, and experience Kakishibu when they visit Yamashiro region. We distributed it to neighboring municipalities, tourist facilities, and educational institutions (Fig. 5.3). To build on this map, we discussed various ideas to incorporate other activities, such as: (1) public engagement activities with young people; (2) setting a walking model course where people

Fig. 5.3 Map of Kakishibu Village Indicating Activities and Things Related to Kakishibu in the Yamasiro region, mainly Kizugawa City

can collect stamps at various locations; (3) having Kakishibu products placed in nearby shrines, temples, cafes, restaurants, inns, hotels, and other places of interest to tourists; (4) development and implementation of Kakishibu hands-on activities; and (5) establishing a Kakishibu Museum. We also made a plan to expand and implement these activities with a cascade pathway: the holding of the National Kakishibu Summit → Holding of the Kakishibu Product Contest → Establishment of the National Kakishibu Promotion Association → Exhibition at the Osaka Expo 2025 → Expansion to Europe and the United States!

However, due to the COVID-19 pandemic, the association had to reduce or cancel many of these planned activities, and this situation remains. The association has received external requests to conduct lectures and workshops on Kakishibu at various places such as elementary schools, community centers, universities, and the Kyoto Prefectural Yamashiro Local History Museum. During the memorable first workshop in 2015, with support from Kyoto Prefecture, we restored a wooden persimmon juice extractor used in the Edo period (Fig. 5.4) and then together with the participants, we actually squeezed astringent persommons to obtain the juice that formed the basis of Kakishibu. In another workshop, we dyed the *washi* scoops used for goldfish scooping with Kakishibu to increase their strength, and had children scoop goldfish. The *washi* paper of the easily torn scoop became very sturdy, and the children were overjoyed to get so many toy goldfish.

Based on these experiences of interacting with the general public, we published the "Keihanna Science Picture Book Kakishibu and Kakitannin" in 2018 and distributed it to various educational institutions. Later, we also published and distributed an English version (2019). In 2021, as part of a collaborative project with the Keihanna Green Innovation Forum, we also published "Kakishibu Dokuhon: the Future of Kakishibu" based on the content of the lectures we had held. However, we did not receive as much interest from young people as we expected even in Japan, and we were not able to disseminate information overseas. In the future, we would like to disseminate information both domestically and internationally using social

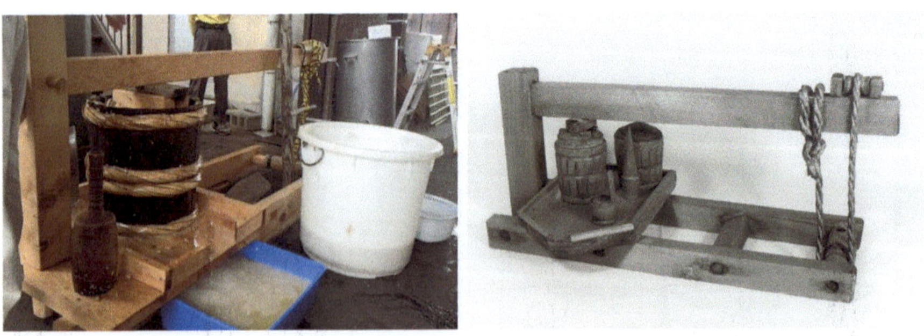

Fig. 5.4 A restored traditional wooden juice extractor used from Edo to Showa period to extract juice from raw persimmons for making Kakishibu (Left: the newly restored version, right: the old picture of traditional original). The left picture is from the author and the black-and-white picture (right) was provided by Mimasu Kashichi Shouten Co., Ltd.

media. In particular, we plan to open a "Kakishibu Channel" on YouTube to increase interest among young people inside and outside Japan, and produce more varied video content.

5.4 Potential for Green Transformations of Kakishibu

Kakishibu is not a versatile industrial material like plastic made from petroleum and coal. It is more expensive, takes more time and effort, and is less durable, but it is a wonderful thing that is environmentally and people friendly. In the future, Kakishibu can open up various possibilities for the advancement of technology and industry in the fields of food, health, and medicine to improve our lives. In addition, various new uses of Kakishibu that are suitable for current society are being considered. What is characteristic of the use of Kakishibu in ancient Japan is that they tried to use it as it is without chemically modifying it excessively. This is different from uses in Western society, where various chemical modifications were made to improve the properties of tree-condensed tannins or to change various properties and create new materials. Here, I would note the development of potential Kakishibu applications informed by a Japanese sensibility, with a focus on technical dimensions:

1. Developing alternatives to disposable plastic containers (various containers, utensils, and tableware found in supermarkets, convenience stores and fast food shops). This involves prioritizing using natural materials as much as possible, such as bamboo, *washi*, and Kakishibu.
2. Developing products that leverage a strong ability to bind with proteins, focusing on minimizing the use of petroleum products and on the protein waste from the medical field, livestock industry, and food-related sectors using Kakishibu.
3. Creating a system to inhibit and inactivate the growth of bacteria and viruses.
4. Developing agricultural pest control agents while reducing the use of chemical agents to limit the chemicals released into the environment.

Therefore, an important area of research is to enhance and strengthen Kakishibu's characteristics by combining it with other natural materials and physical processes such as mixing, bonding, and adhesion as a green transformation of Kakishibu. In particular, it is believed that combinations with various proteins (including fibrous proteins), natural oils, polysaccharides (viscous HMW substances), safe minerals, among others, is promising. As traditional techniques, the production methods of Japanese *Ise Katagami* (Fig. 5.5), *hakuutishi* papers used for the hammering process of gold leaf, paper clothing, *bangasa* (Japanese paper umbrellas) and wood paint such as *shibuzumi* and *bengara* painting are one of good examples. In *Ise katagami* and *bangasa*, plant oils and natural resins are used in addition to *washi* (Japanese paper) and Kakishibu, and in the case of *hakuutishi*, the protein of egg whites is used with *washi* and Kakishibu. Paints made by mixing Kakishibu with soot (formed when pine trees are burned) and with *bengara* are used as a durable paint of woods.

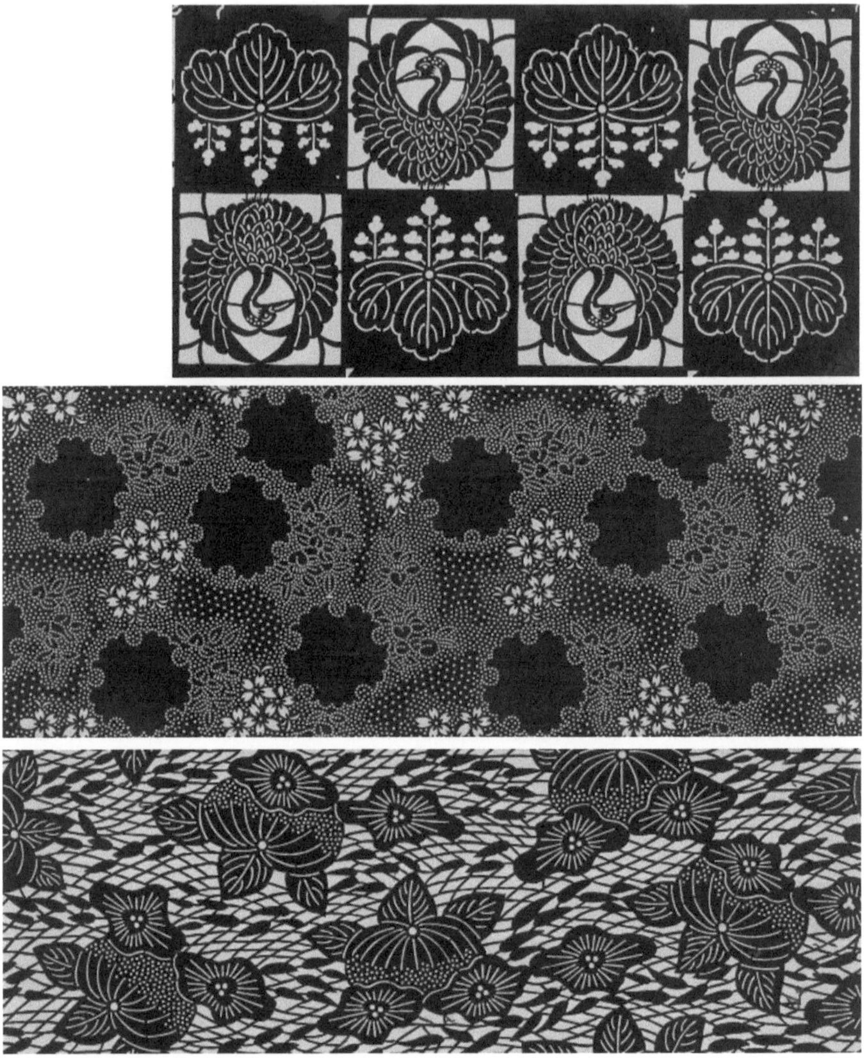

Fig. 5.5 *Isekatagami*, a traditional art craft made from Kakishibu and *washi*. Top: *Kiri ni Tssuru*, Middle: *Yukiwa ni Sakura,* Bottom: *Hana Tachibana.* All images were downloaded and remixed from Minna no Kyodo Shiryo Shitsu Digital Archive, Bunkazai Ka, Suzuka City (CC-BY-SA). *Isekatagami* is made of Mino *washi*, which is laminated and reinforced with Kakishibu as a backing paper, and on which artisans carve designs by hand using one of engraving knifes. These unique patterns are painstakingly created by skilled craftsmen using a variety of carving knives and techniques. This katagami is mainly used to dye patterns and designs for yuzen, komon, yukata, and other kimonos. Recently, they are also used for interior decorations such as sliding doors and shoji screens, and LED lighting

Japanese paper originally uses sticky polysaccharides of *Tororoaoi* (*Abelmoschus manihot*) for fiber bonding.

The green transformation of Kakishibu, based on fusion with natural materials, and building a new natural material culture will contribute to replacing the currently dominant plastic culture. Currently, other effective higher polymer natural materials such as starch and cellulose, which have a long history of research, are already on the market as materials with various different characteristics achieved through partially chemically modifying them. Rather than those that are overly modified, we would like to draw attention to those that retain high biodegradability, such as different material combinations with Kakishibu for the creation of new materials. Chitin and chitosan extracted from the discarded shells and limbs of shrimp and crabs, fibrous proteins such as collagen and peptides from discarded animal skins and hair, viscous polysaccharides from konjac and seaweed, and various plant oils are expected to be good material partners for Kakishibu. Today, technologies have been developed that utilize enzymatic reactions as an eco-friendly chemical reaction. By selecting enzymes that are not denatured by tannins, there is the possibility of combining kakitannins with other natural materials to produce new natural materials.

In the future, it will be necessary to promote demand for Kakishibu products, design these products, build a mass production system, and then implement price setting and container and packaging design. It is crucial to offer products that surpass the drawbacks of natural materials for general consumers or to provide product education. A strategy to generate a big swell of consumption through publicity and dissemination throughout society is needed.

5.5 New Frontiers in the Kakishibu Industry

Traditional daily life before the Industrial Revolution was not easy. It was laborious and extremely inefficient compared to our current lifestyle. But were people unhappy? Since then, people have been striving for economic improvement and convenience, but what has this resulted in? From some, aren't these efforts causing damage to the global environment and the human spirit? Does constantly working to buy the latest expensive smartphones constitute a rich life? Recently, there have been many such criticisms, prompting social reflection. Even if it takes some effort and is economically inefficient, a life that is kind to people and the environment, and is slow and peaceful, may lead to happiness.

In Japan, Kakishibu has a long history, and its main component, kakitannin, is a very unique water-soluble natural polymer compound. In recent years, there are questions about what paths the declining Kakishibu industry will take in the future. How may it be possible to create opportunities for a V-shaped recovery? If social pressure is applied in the right direction, a new path may emerge, driven by new social demands. These demands could include legal regulations and implementation related to the production and use of plastic, or if multiple local governments invest in the formation of an eco-society that prioritizes natural materials and takes concrete action to aim for a society that is free from oil and plastic. These could be big steps toward major change.

In order to revive the Kakishibu industry, I would like to envision a model society that incorporates the essence of traditional Kakishibu culture into contemporary society. I also want to highlight the significance and role of small- and medium-sized regional cities in modern society, especially in relation to large cities (see Figs. 5.6 and 5.7). Large cities have concentrated populations, a large number of cars, less green areas, and cramped living environments due to the small amount of space. Furthermore, large amounts of energy are consumed. As a result, the air is polluted, high volumes of waste (including a lot of plastic waste) are produced, and there is often less interaction and contact between individuals and communities, leading to a lack of emotional fulfillment for people. However, there is an abundance of human resources, money, and information, and it is stimulating.

I would like to consider a system in which "small towns" can provide what is lacking in cities and meet social needs. In a small town not far from large cities, we aim to create a "Kakishibu Hamlet." At the core of this aim is to create a society free of oil and plastic, and to prioritize the use of natural materials. In the Kakishibu Hamlet, we will set up a "highly planned diversified persimmon orchard" that also serves as a deterrent against harmful animals, making effective use of abandoned bamboo forests and farmland. In this orchard, we will build a system that introduces the latest weed control, irrigation, and immature persimmon harvesting techniques,

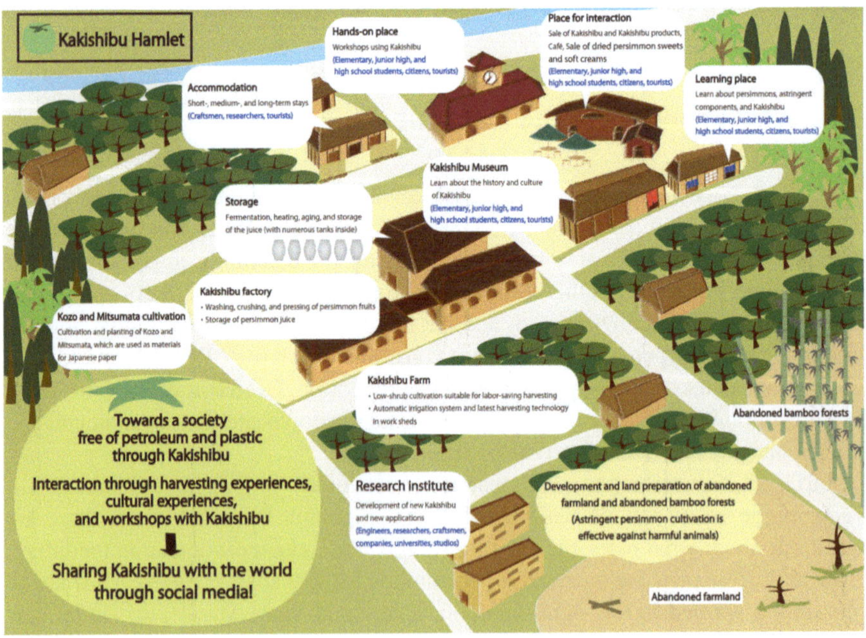

Fig. 5.6 Diagrammatic representation of the components of Kakishibu Hamlet

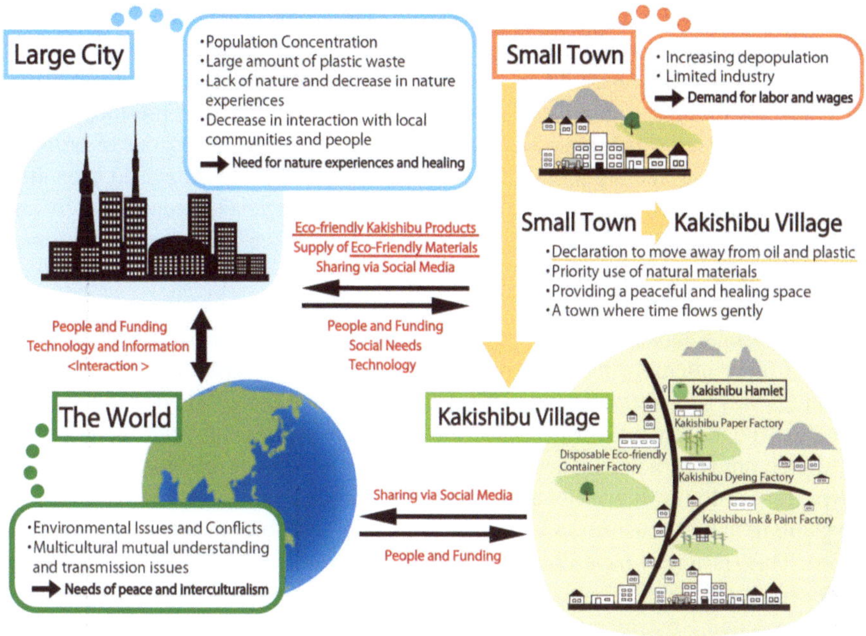

Fig. 5.7 Proposal for a slow life and the concept of a Kakishibu Village, including Kakishibu Hamlet, promoting organic interaction between small- and medium-sized regional cities and large cities

based on a comprehensive design that prioritizes labor-saving. We will use organic farming methods that use goats and deer for weed control and natural compost. The immature green persimmons harvested in midsummer would be squeezed in this village, followed by fermentation and aging to produce Kakishibu.

Around this "Kakishibu Hamlet," we will build and place a factory for producing *shibugami,* Kakishibu-treated papers, using the produced Kakishibu, a workshop for Kakishibu dyeing, a factory for producing *shibuzumi*, and disposable eco-containers and tools. We will also create bamboo groves and gardens for Mitsumata and/or Kozo, paper bush used as raw material of Japanese paper, to obtain the raw materials for washi, and place them in the surrounding area. We will design a distinctive regional community by setting up the entire area as a "Kakishibu Village." From the small town, including this Kakishibu Hamlet and Kakishibu Village, we will supply eco-materials combined with eco-Kakishibu products and other natural materials to large cities. In addition to how to use and maintain these materials, we will provide ways to live a new life centered on nature and natural materials. At the same time, we will disseminate this information to the world through social media. In addition to things, small- and medium-sized regional cities can offer opportunities for experiential learning of the value of casual, spacious, and quiet natural spaces, casual interactions between people, and warm conversations as mental healthcare benefits for people who are exhausted from large cities (Fig. 5.7).

The rapid development of the leather and tannin industries in medieval Europe and the rapid decline of Kakishibu culture during the Showa period in Japan, due to the influx of plastics, share a common social upheaval. The origins of these social changes are thought to have been greatly influenced by the wartime idea of efficiency first, which was far removed from the consciousness and actions of ordinary citizens. However, the big problem is that this efficiency first was valued even after war, and it continued to accelerate. In order for Japan to develop economically and become larger, it was considered good to develop rapidly, and this era has continued for a long time. Is the current society that has been created as a result what we have been seeking?

In traditional Japanese daily life, tea leaves were hand-picked from vast tea fields using a labor-intensive method and then steamed and hand-rolled. The resulting green tea was served in a tatami-matted Japanese-style room, where one would sit and pour the tea into a favorite teapot, then into a favorite teacup, one cup at a time. It was common to admire each vessel, enjoy the aroma and color of the tea, and savor the tea slowly. It was not about drinking, but savoring. This process was time-consuming and expensive, and washing the teapot and teacups also took time and effort. This spirit may have something in common with savoring wine at a medieval European dining table. However, nowadays, many young people simply buy bottled tea from vending machines when they want to drink green tea, and drink it casually while walking or sitting on a bench in a park or on the roadside. They drink not to savor, but to quench their thirst. The tea they drink is mass-produced in vast tea fields that have been efficiently cultivated, where almost all persimmon trees in tea fields have been cut down and the tea leaves have been mechanically picked. The cost is extremely low, it requires no effort, and it can be stored. Above all, it is cheap and easy to drink, and the cleanup is simple. Often, you see empty plastic bottles scattered on the roadside or in parks. Which lifestyle would you choose? Is this the affluent life we have been longing for in Japan?

Finally, I would like to leave readers with some reflections. In current society, how do we recognize traditional knowledge, understand its importance and significance, and connect it to action? How can we prioritize natural materials in the future, distribute them throughout modern society, and coexist with synthetic chemicals? It is difficult to create big social changes just through the activities of scattered groups, as these efforts will be piecemeal and short-lived. In this convenient and, in a sense, comfortable modern society, what brings about collective change consciousness and action in society? The key lies in the lifestyles of an era when sustainable societies and circular economies existed, built upon the use of Kakishibu. I do not mean to suggest we should return to that era. Rather, I believe that by reviving the use of Kakishibu, we can reclaim the mindset of that time; nature as a being to coexist alongside of. By harmonizing Kakishibu and the mindset it evokes with modern life, we can create new sustainable societies and circular economies.

References

Fujishima, H., Tijiwa, H., Shiraishi, M., Ushijima, K., & Matsuda, K. (2011). Kaki 'Fuyu' no cho-teijuko ichimonji-seishi ga sagyosei, shuryo, kajitsu-hinshitsu ni oyobosu eikyo (Effects of ultra-low tree height single-line pruning of persimmon 'Fuyu' on workability, yield, and fruit quality). *Fukuokaken-Nogyosogosikenjo-Kenkyuhokoku (Fukuoka Prefecture Agricultural Research Center Research Report), 30*, 48–55.

Hamazaki, S. (2010), Ethanol de datsuju shita kajitsu o mochiita kakitannin no jinsokuna chōsei-hō (Quick preparation of kakitannin fromde-astringent persimmon fruit treated by ethanol) *Horticultural Research in Japan, 9*(3), 367–372.

Iwamoto, M. (2013). Seichōzai no rekishi-teki haikei to seishu e no ōyō (Historical background of the finning agent, kakishibu and its application to sake). *Journal of the Brewing Society of Japan, 108*(5), 318–325.

Japan Fruit Association. (2022). *Sekai no Kajyu Seisan Gaikyo 2021 nen ban (World Major Fruit Production Overview 2021 Edition)*

Ministry of Agriculture, Forestry and Fisheries. (2019). *Fruit production and shipment statistics (persimmons).* Retrieved September 16, 2025, from https://www.e-stat.go.jp/dbview?sid=0003424060

Ministry of Agriculture, Forestry and Fisheries. (2020). *Fruit production and shipment statistics (persimmons).* Retrieved September 16, 2025, from https://www.e-stat.go.jp/dbview?sid=0001931838

Ministry of Agriculture, Forestry and Fisheries. (2021). *Fruit production and shipment statistics (persimmons).* Retrieved September 16, 2025, from https://www.e-stat.go.jp/dbview?sid=0002018859

Ministry of Agriculture, Forestry and Fisheries. (2022). *Fruit production and shipment statistics (persimmons).* Retrieved September 16, 2025, from https://www.e-stat.go.jp/dbview?sid=0002117465

Ministry of Agriculture, Forestry and Fisheries. (2023). *Fruit production and shipment statistics (persimmons).* Retrieved September 16, 2025, from https://www.e-stat.go.jp/dbview?sid=0002119821

Naito, T., Yamamura, H., & Miga, T. (1975). Saijokaki no yakuzaitekika ni okeru NAA no sanpujiki, nodo, oyobi sonotano shoyoin ni tsuite (Effectiveness of NAA spray in thinning of Kaki 'Saijo' fruits as related to application time, concentration and other factors. *Bulletin of the Faculty of Agriculture, Shimane University, 7*, 1–8.

Neto, R. T., Santos, S. A. O., Oliveira, J., & Silvestre, A. J. D. (2020). Biorefinery of high polymerization degree proanthocyanidins in the context of circular economy. *Industrial Crops and Products, 151*, 112450. https://doi.org/10.1016/j.indcrop.2020.112450

Oda, K. (2014). Kakishibu kamikowashi no sozai kenkyu oyobi fukushoku design (Materials research on Kakishibu-Kamiko and its uses in dress design). *Journal of the Japan Research Association for Textile End-Users, 55*(5), 349–355.

Pizzi, A. (2019). Tannins: Prospectives and actual industrial applications. *Biomolecules, 9*, 344–374. https://doi.org/10.3390/biom9080344

Sakate, Y., & Kojima, N. (2015). Washi no dento-teki hassui bosui kako no saigen - Dento-teki washi bunka no saiko ni mukete (The experiment to make a waterproofed "washi" on a traditional technique – Toward a revival the obsolete culture of the washi, traditional Japanese handmade paper. *Bulletin of Japanese Society for the Science of Design, 62*(6), 43–50.

Singh, P. A., & Kumar, S. (2020). Applications of tannins in industry. *IntechOpen.* https://doi.org/10.5772/intechopen.85984

Seino, H., Akashi, M, Nonaka, M., & Ishigaki, R. (2016). *Kakikako yo Kajitu no shoryokuteki na saibai gjuysu (Labor saving cultivation technology for persimmons for processing).* Tohoku Nougyo Kenkyu (Tohoku Agriculture Research), 69, 79-80.

Yamamura, H., & Naito, T. (1975). Kaki ni okeru NAA no tekikakikko ni tsuite (dai 1 po) NAA shyori ni yoru rakka to kajitu no naiseiseityosetubussitu tono kankei (Mechanism of the thinning action of NAA in kaki fruits: I. Relation between NAA-induced fruit abscission and endogenous growth substances in fruit tissues. *Journal of the Japanese Society for Horticultural Science, 43*, 406–414.

Tomoaki Matsuo is an emeritus professor at Kagoshima University in Kagoshima City, Japan. He earned his Doctor of Agriculture degree from Kyushu University's Faculty of Agriculture for his research achievements, which included analyzing the chemical structures of astringent components in astringent persimmons and condensed tannins in other fruits. Following his retirement, he established the Scientific Research Association for Kakishibu and Kakitannin. He routinely conducts Kakishibu lectures and exhibitions, organizes workshops, maintains a website, and publishes the Kakishibu Newsletters. Through these activities, he disseminates information regarding "the wonders and appeal of Kakishibu, and its industrial applications" both domestically and internationally. Concurrently, he is involved in technological development aimed at new applications of Kakishibu.

Chapter 6
Socio-technical Experiments with Kakishibu: Student-Led Innovation with Businesses

Toshihiko Matsuda

Abstract This chapter discusses a case study of inquiry-based learning in Japanese high schools, where students used Kakishibu (persimmon tannin), a local resource, to address the challenge of creating a more sustainable society. Kakishibu is a traditional Japanese material whose value has been largely overlooked in recent years. However, by exploring solutions to environmental issues such as microplastics and per- and polyfluoroalkyl substances (PFAS), the students focused on the functional properties of Kakishibu, including its waterproofing, antibacterial, and deodorizing qualities. They took on the challenge of developing multiple Kakishibu-based products aimed at reducing plastic use, including paper bags, cosmetics, agricultural mulch, and food packaging paper. Through collaboration with companies, universities, and local governments, the students gained real-world developed, worked on business plan proposals and delivered presentations, and demonstrated flexibility and initiative without fear of failure. In doing so, they proposed innovations that could revive Kakishibu, a local speciality, for broader use in society. This initiative exemplifies a collaborative model that connects schools, communities, and businesses and research institutions, demonstrating that high school students' learning can drive societal change. The challenges and initiatives undertaken in an environment that allows for failure highlight the potential to rediscover the value of local resources and contribute to building a sustainable society through education.

Keywords Inquiry-based learning · Local resources · Sustainability · Student-centered learning · Social issue resolution · Environmental education · Plastic reduction · Corporate collaboration · Traditional technology · Innovation

---

T. Matsuda (✉)
Ikubunkan Institute of Education, Bunkyo City, Tokyo, Japan
e-mail: t-matsuda@ikubunkan.jp

K. Kano et al. (eds.), *Traditional Knowledge for Green Innovation with Japanese Persimmon Tannin, Kakishibu*, https://doi.org/10.1007/978-981-95-7805-4_6

6.1 Current State of Japanese High Schools and New Initiatives

I was appointed as an agricultural teacher at high schools by the Kyoto Prefectural Board of Education in 1993. In Japanese high schools, regular transfers are common, and teachers typically work at multiple schools over the course of their careers. I would like to begin with the story of Kyoto Prefectural Katsura High School, where I was assigned from 2009 to 2018, which led me to meet Professor Emeritus Tomoaki Matsuo, one of the contributors to this book, and then move on to the story of Kizu High School, where we launched a full-scale initiative on Kakishibu.

At the time of my assignment to Kyoto Prefectural Katsura High School, Japanese high schools traditionally emphasized a curriculum oriented toward university entrance exams, focusing on the acquisition of knowledge. Consequently, student evaluations were also knowledge-based, and teachers were reluctant to collaborate with the outside world and disseminate information outside the school (Japan Science and Technology Agency, 2009). Katsura High School was no exception, despite being selected as a Super Science High School. A Super Science High School is an initiative to "cultivate students' scientific inquiry skills through advanced science and technology, and science and mathematics education (hereinafter referred to as 'science and mathematics education') in high schools and integrated junior and senior high schools (including integrated and cooperative junior and senior high schools), and thereby nurture future internationally active science and technology personnel" (Japan Science and Technology Agency, 2021).

To overcome these challenges, Kyoto Prefectural Katsura High School decided to emphasize inquiry and engage in activities to solve real-world problems. Specifically, we started an initiative involving a seed bank for *kyo-yasai*, Kyoto traditional vegetables, with the students. Kyoto is far from the sea, making it difficult to obtain seafood. As such, vegetables have become the center of diets in Kyoto. Vegetables have been gathered from all over the country as tribute goods since Kyoto flourished as a capital more than 1,000 years ago. These vegetables, nurtured by Kyoto's unique climate, fertile soil, and abundant water, and improved by farmers' techniques, mark the beginning of *kyo-yasai* (Takashima, 2003). They have also supported Kyoto's food culture, such as *syojin-ryori*, Buddhist Vegetarian Cuisine, and *obanzai*, a traditional style of cooking specific to Kyoto, contributing to *washoku*, Japanese cuisine, being registered as a UNESCO Intangible Cultural Heritage in 2013 (Namimatsu, 2020). Seed banks could play a role in maintaining access to a range of *kyo-yasai*. A seed bank is a facility or organization where plant seeds are stored and managed through human labor. They collect the seeds of a wide variety of plants around the world and store them so that the species, and their genetic resources, can be preserved and used by future generations even if they are at risk of extinction in their habitats (Nishikawa & Hamaguchi, 2018).

However, activities such as seed banks were not included in the regular curriculum of the school at that time. The usual curriculum of the general course in Japan, which is differentiated from the specialized course and integrated course and consists of 74% of high school students (Ministry of Education, Culture, Sports, Science and Technology, 2024), is usually composed of mathematics, science, Japanese, social studies, English, arts (music, art, calligraphy), and physical education (Ministry of Education, Culture, Sports, Science and Technology, 2009). Katsura High School was a bit special because it was agricultural, with subjects and research projects to learn cultivation techniques for vegetables, flowers, and other plants. There was a research project as part of the specialized course, but it was formalized and the content was more work-like than inquiry. In the latest curriculum, there are subjects that emphasize inquiry, such as science and mathematics (Ministry of Education, Culture, Sports, Science and Technology, 2018). Therefore, it may be possible to incorporate activities oriented toward real-world problems within the curriculum. However, it seems difficult to carry out all the initiatives and connect with various stakeholders within the existing curriculum. Therefore, the content I will now introduce is from activities that have been conducted outside the curriculum, but within the school's purview. These activities are therefore distinct from other extracurricular activities in Japan called club activities, which are popular and where teachers are involved as advisors (Ministry of Education, Culture, Sports, Science and Technology, 1978).

Currently, the vegetables circulating in the market are mainly easy-to-eat varieties with no peculiarities. However, traditional vegetables often have unique aromas and bitterness, and recent studies have revealed that they contain components with cancer-suppressing and antioxidant effects (Nakamura et al., 2004). Therefore, we launched a project with students to brand *kyo-yasai* as "functional vegetables" and proposed new processing methods to expand demand. As a result, we were able to develop products in collaboration with many companies, and demand increased.

For example, one of the *kyo-yasai*, *katsura-uri* (Japan's heirloom pickling melon, Cucumis melo var. conomon) was cultivated on about 30 hectares starting from around 1942 as a high-quality ingredient for pickles. However, during the war, the cultivation of starch crops was enforced, and planting was discontinued. After the Second World War, the demand for pickles decreased due to changes in dietary habits, and cultivation continued to decrease. The number of farmers cultivating it eventually decreased to one, entrusted by Kyoto City government to preserve the seeds.

High school students who learned of this situation considered ways to expand demand even with current dietary habits and focused on emerging functionalities of Katsura-uri. This included research on the cancer-suppressing effects of the aroma components (MTPE, etc.) contained in *katsura-uri* (Nakamura et al., 2008), and the fact that *Katsura-uri* is characterized by a fragrance similar to muskmelon while containing almost no sugar. Therefore, sweets were developed for diabetics by adding sweetness with artificial sweeteners. This product received a great response, and several farmers started cultivating it, inspired by the students' efforts.

Furthermore, local governments, universities, and companies in Kyoto Prefecture established the "Kyoto Vegetable Functional Utilization Promotion Liaison Council" in 2013 to promote the spread of *kyo-yasai*, and high school students were also asked to participate in their activities. In this way, we realized that the activities of high school students could be a driving force for these initiatives and have broader impacts on society.

Through the activities of these high school students, we were able to build many communities. For example, during this time, Professor Koshi Nakamura (Life and Environmental Engineering) of Kyoto Prefectural University, who was conducting joint research on *kyo-yasai*, introduced us to Professor Emeritus Tomoaki Matsuo, a leading researcher on Kakishibu at Kagoshima University and one of the authors of this book. In this way, through my involvement with *kyo-yasai*, I witnessed high school students confronting social issues, taking action, and working through various regulations and constraints with the cooperation of administrations, companies, and research institutions.

6.2 Starting Point of Working on Kakishibu

I started working on Kakishibu when I was transferred to Kyoto Prefectural Kizu High School in 2018 by the board of education. In Japan, teachers are transferred to different schools roughly every 10 years. I had already spent 10 years at Katsura High School, so I was transferred to Kizu High School. In the Yamashiro region of Kyoto Prefecture, where Kizu High School is located, Kakishibu was once utilized as a local resource. However, with societal changes, its presence gradually faded, and it was on the brink of disappearing from the area. Therefore, I participated in a Kakishibu research meeting held locally and reunited with Professor Emeritus Matsuo, who is also a director general of the Scientific Research Association for Kakishibu and Kakitannin. At that meeting, I was introduced to Mr. Takeo Mimasu, who is the chairman of a Kakishibu factory known as Mimasu Kashichi Shouten Co., Ltd. and an author of this book. Matsuo and Mimasu had been working hard to promote Kakishibu by launching various events and research meetings. They had also proposed a joint research project on Kakishibu to Kizu High School before I was transferred to the school. However, the school was reluctant to incorporate a new program due to the daily educational activities, and the collaboration did not progress as expected.

Soon after the meeting with Matsuo and Mimasu, I called on the students and decided to set up a research team. However, as mentioned earlier, under the Japanese national curriculum, initially there were hardly any students interested in research activities. I started the activity with just one student from my class. To start the activities in earnest, we first visited Mimasu to learn about Kakishibu. Despite being 86 years old at the time, his passion for Kakishibu was stronger than anyone else's, and he enthusiastically spoke to the students. To increase the production of Kakishibu, he had planted *Tenno-gaki* (Tenno persimmons), the raw material, not only in Kyoto Prefecture but also in other prefectures. He also developed and sold a variety of products such as bed covers, pillow covers, and mats under the brand "Kakishibu House Mimasu." Mimasu also shared various stories with the students. Hearing about the various functionalities of Kakishibu, its history, and his commitment to continue protecting it while the number of local dealers handling Kakishibu continued to decrease, seemed to strengthen their desire to convey the charm of Kakishibu to many people.

Afterwards, I realized that it was difficult to set up the equipment and chemically analyze the various functionalities of Kakishibu, such as antibacterial, insect repellent, and waterproof properties, in the school environment. Therefore, we decided to develop new ways to utilize Kakishibu and propose them to companies and various institutions. As much as possible, the role of considering and verifying the utilization methods, and approaching companies and various institutions, was undertaken by the students, and I took on the role of advising the project, public relations to the media, managing activity expenses, and selecting business contest entries, which are described in the following sections.

6.3 Increasing Awareness of the Issue of Plastic Products in Japanese Society: Trends Toward Charging for Plastic Shopping Bags

I launched the Kakishibu project with three students. First, I approached a student, Yuko Chikami, from the first-year high school class I was responsible for. I chose her because she had the best grades in the class and was not a member of any after-school club activities. When I asked her why she didn't join a club, she said, "I want to go home early." At first, she was not very enthusiastic, but she reluctantly agreed to participate. However, she seemed to have difficulty getting motivated on her own. Second, I approached another student, Kazusa Ito, from the Kendo club I was advising, and she was more positive, saying, "I don't mind participating." Finally, there was a student, Yuri Takaya, in the class who wanted to go on to study agriculture, so I also approached her, saying, "It might be useful for your future studies, why don't you participate?" She also showed interest, saying, "I don't mind trying."

In the process of planning a product using Kakishibu, we identified the low awareness of Kakishibu in Japan as a significant challenge. This was also acknowledged during a visit to the shop of Chairman Mimasu. Currently, products made from Kakishibu are mainly used in kimonos and traditional crafts that utilize Kakishibu dyeing, so there were hardly any opportunities for the general public to handle them. In fact, many students said, "I have never seen a product using Kakishibu before." Therefore, we decided to consider ideas for applying Kakishibu to everyday products. Around that time, the issue of marine pollution caused by microplastics was increasingly reported as a major social problem in Japan. A television program featured the severe environmental pollution on Boracay Island, a tourist destination in the Philippines. The island was closed for 6 months for pollution control measures, and although tourism resumed afterwards, the use of disposable plastics such as shopping bags and straws was prohibited (Association of Marine Leisure Promotion, 2023). One scene in the program left a strong impression. It was when a paper bag filled with several layers of raw fish broke due to moisture.

At that moment, I had an idea, "If we apply Kakishibu to the paper bag, it could improve its water resistance and solve the problem." In Japan, the government had been working toward charging for shopping bags since around 2019, which was eventually implemented on July 1, 2020, under the Law for the Promotion of Sorted Collection and Recycling of Containers and Packaging. However, the global trend is to ban the use of plastic shopping bags altogether, and I felt that Japan's charging measures were still insufficient. Therefore, I proposed to the students that we strengthen paper bags with Kakishibu and therefore try to replace regular shopping bags, often made from paper. Kakishibu Paper bags are made of natural materials and eventually decompose in nature, which is an advantage in terms of low environmental impact. However, they are weak against water and have durability issues. Some products have improved water resistance because the inside of the paper bag is coated with plastic, but this makes it difficult to separate the plastic and paper parts, making it impossible to recycle and ultimately increasing burden on the environment. Therefore, we focused on the water resistance, durability, and the deodorizing properties that Kakishibu provides. Kakishibu is a natural material and eventually decomposes, so we thought it could improve the performance of paper bags without burdening the environment, and decided to proceed with exploration and practice.

6.4 Kakishibu Paper Bag Development Project

Students (Chikami, Ito, and Takaya) started experiments to confirm whether Kakishibu could enhance the durability and water-repellent properties of paper bags. To demonstrate the performance of Kakishibu, they used newspaper, a familiar material, as a test material. In Japan, over 96% of the daily circulation (The Japan Newspaper Publishers & Editors Association, 2024a) of 25 million daily newspapers (The Japan Newspaper Publishers & Editors Association, 2024b) is delivered daily to individual households, making it a familiar presence in Japanese life. The raw materials for newspaper are mainly thermo mechanical pulp (TMP) and wastepaper. Because it is a highly absorbent material, it easily tears when wet.

The test samples were compared with untreated samples. The test samples were treated with Kakishibu diluted twice and three times with water (Fig. 6.1). All application of Kakishibu on newspaper was done by hand. The newspaper was prone to tearing, and even moving it to dry could cause it to tear, so each newspaper was carefully painted with a brush. Afterwards, it takes about 24 hours to dry indoors, and a large space is needed to dry the many newspapers painted with Kakishibu. Therefore, we utilized classrooms that were not used on weekends and dried them on holidays. Durability was tested by hanging a 2.5-liter plastic bottle (2-liter plus 0.5-liter bottles) filled with water in the bag, and water resistance was tested by

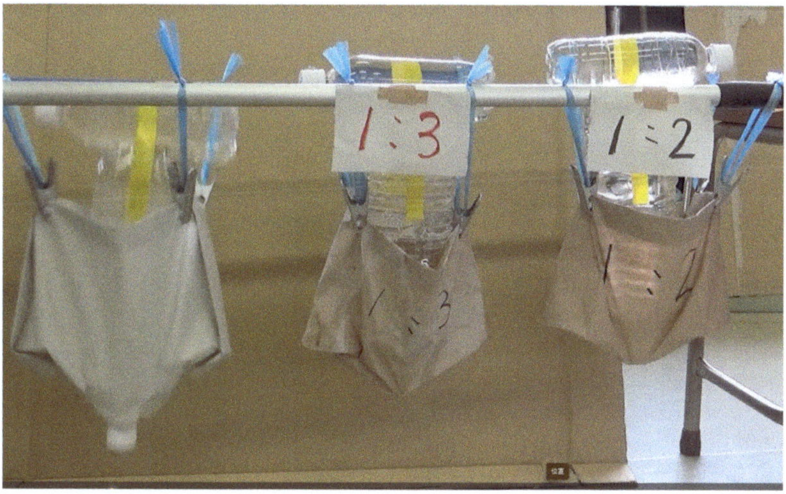

Fig. 6.1 Verification of improvements to durability through applying Kakishibu paint on paper made of the same material as newspaper

wetting the bag with water and filling it with raw konjac (a type of plant) and a weight. Konjac was used because it has the property of retaining moisture for a long time.

In Fig. 6.1, the photo on the left shows the untreated area, the middle (marked as 1:3) is a ratio of 1-part Kakishibu to 3 parts water, and the right (marked as 1:2) is a ratio of 1-part Kakishibu to 2 parts water. We hung them inside bags made from 500 ml plastic bottles. The untreated newspaper tore immediately within 2–3 seconds, but the areas treated with Kakishibu did not tear even after more than 2 hours, and the experiment was concluded at that point as no tearing seemed imminent.

This result confirmed that newspaper treated with Kakishibu significantly improves in durability and water resistance compared to untreated ones. Overall, these experiments were able to demonstrate that if used to produce a plastic bag substitute for paper bags, Kakishibu would provide sufficient durability and water resistance, as well as the added value of odor prevention.

However, we felt that the fact that it takes about 24 hours to dry indoors could be a significant challenge for widespread adoption. While we confirmed that applying Kakishibu improves durability and water resistance, quantification based on industrial standards is necessary to distribute it as a commercial product. Prototyping went well, but it became clear that there are several challenges to distribute it in the broader market. To solve these challenges, we decided to look for companies that would consider an industrialization system to mass-produce paper bags coated with Kakishibu.

6.5 Collaboration with Universities and Companies

The students, aiming for the industrialization of paper bags coated with Kakishibu, visited a local paper product manufacturer in 2018 to seek their opinion. The company showed interest, but pointed out challenges such as the time-consuming drying process, increased cost due to Kakishibu coating, and uncertainty of demand. In particular, the lack of precedent was identified as the biggest hurdle. In order to address some of these challenges, several partnerships were pursued.

Firstly, in order to increase the appeal of Kakishibu with more people, we consulted with Kano at Shiga University, who had helped us with developing previous workshops using *kyo-yasai* and Japanese tea and one of the editors and authors of this book. As a result, he agreed to develop an educational program in the form of a workshop utilizing Kakishibu.

Additionally, I asked media personnel to directly interview the students about their project to publicize paper products using Kakishibu. As a result, we were able to get featured in a local newspaper. This coverage led to contact from and dialogue with Goto at Maxell, Ltd., who is one of the authors of this book. The dialogue prompted us to initiate cooperation with commercial and industry partners.

First, we made a proposal demonstrating the potential of paper products, including Kakishibu paper bags, and sent it with a letter to the chairman of one of the world's leading fast-fashion manufacturers representing Japan. We approached this company because we learned that they were reducing disposable plastic packaging in stores by 85% and gradually switching to environmentally friendly paper bags using recycled paper in all their stores worldwide. As such, we anticipated that they might be interested in our work.

The initial response was "interesting, but difficult to adopt in our company." However, a week later, we received a call from the secretary of the company's executive officer, expressing a desire to hear more about it. When the high school students explained the characteristics and potential of Kakishibu, they responded, "Very interesting. We don't know if it can be realized immediately, but we will consider it internally." During subsequent meetings, we learned that the company was facing a problem with recently introduced paper bags tearing in the rain, leading to complaints being posted and spread on social media.

Furthermore, we were invited to the Tokyo headquarters and had the opportunity to exchange opinions with the management executives. It was decided to set up a Kakishibu PR booth at the Kyoto store and conduct a trial service offering paper bags coated with Kakishibu. However, this plan did not come to fruition as all events were canceled due to the spread of COVID-19 shortly thereafter.

In addition to this, we also approached a retail company that operates hundreds of stores worldwide, and they showed interest. This company is committed to selecting materials that consider the global environment through eliminating waste in all processes and a commitment to providing customers with what they truly need in the form they truly need. The company's representative explained that they "want to switch not only paper bags but all packaging products from plastic to recyclable paper products." However, our collaboration with this company was also halted due to the spread of COVID-19.

Through these activities, we received a strong, enthusiastic response when we conveyed the appeal of Kakishibu to various companies. However, high school students have a limited time to engage in these activities, and the impact of COVID-19 imposed significant constraints on their efforts.

6.6 Challenge and Impact of the Contest

The students believed that to expand their sales channels, they needed to approach companies and convey the appeal of Kakishibu to a wider audience. To achieve this, they decided to participate in various business contests, thereby spreading the appeal of Kakishibu and demonstrating its potential as a viable business opportunity. The reason for focusing on business is that they believed that if Kakishibu was only protected as a traditional resource through volunteer activities, it would eventually decline through uneven resourcing constraints. This has also been a consistent concern since the initiative with *kyo-yasai*.

To achieve our business plan of replacing disposable plastic products circulating in the market with paper products coated with Kakishibu, material constraints need to be overcome. As the amount of disposable plastic products currently in circulation is very large, the realization of our business plan requires an increase in the production of "astringent persimmons," the raw material for Kakishibu. Therefore, we thought about the use of abandoned farmland in mountainous areas in Japan, which is becoming a serious problem, to help address this problem.

In Japan's mountainous regions, the number of abandoned farmlands is increasing year by year due to depopulation and damage caused by wild animals such as monkeys, deer, and wild boars (Sakata, 2010). These neglected forests in Japan are an ongoing problem, and proper logging and consumption of wood are required (Kobayashi & Higano, 2000). We thought that cultivating astringent persimmons on this land could contribute to revitalizing these regions. Astringent persimmons are resistant to damage from wild animals due to their strong astringency and can be cultivated even on slopes. In addition, the roots of the trees stabilize the soil, contributing to the prevention of landslides. Because astringent persimmons are not used for food, they also do not require management such as fruit picking and pesticide spraying, which is a significant advantage. We hope that by increasing the demand for paper through this business, we can also promote the proper use of forest resources.

Furthermore, while there may be potential competitor products, these also have disadvantages. For example, the raw material for biodegradable plastics, starch, is derived from grains, and there is a concern that it could lead to a food crisis (Munesue & Masui, 2012) as the world population is predicted to reach a peak of around 10.3 billion people in the mid-2080s (United Nations, 2024).

In this regard, paper products using Kakishibu could be a sustainable alternative while reducing environmental impact and pressure on existing agricultural land.

The following business plan of Kakishibu paper bags that students proposed at business contests incorporates these arguments.

6.6.1 Features

Kakishibu has not only strength but several other useful properties, for example, water-resistant, insect-repellent, and deodorizing effects. It was formerly used as a material to enhance the durability of items such as Japanese umbrellas and furniture, and since it is a natural material, it does not burden the environment.

6.6.2 Novelty and Competing Products

Similar technologies include biodegradable plastics and biomass plastics. However, biodegradable plastics pose a significant problem in terms of marine pollution, as not all biodegradable plastics are difficult to decompose in aquatic environments (Mitsubishi Research Institute, 2019). Biomass plastics are considered environmentally friendly, but they are expensive as they are made from starch, and there are limitations to their widespread use considering the potential for exacerbating global food crises and events like the Amazon forest fires caused by the expansion of farmland for increased grain production (Yoshikawa, 2015). Moreover, the French government has also set a target to reduce the annual market input of disposable plastic packaging, including those made from biomass and biodegradable plastics, by 20% from 2018 levels by the end of 2025, with at least 50% of the reduction to be achieved through reuse and recycling (Yamasaki & Watanabe, 2024). Regarding Kakishibu, while the price of Kakishibu may be high in the start-up phase of the business, we believe that costs will decrease if we can produce a large amount of Kakishibu at a low cost.

6.6.3 Strengths in Business Development

Currently, many companies are working to solve environmental problems, and we believe that the introduction of Kakishibu paper bags will not only be good for the environment but also enhance the overall image of the company as working with a unique natural material from Japan. In fact, major apparel companies have declared that they will reduce disposable plastic packaging in our stores by 85% and gradually switch shopping bags in stores worldwide to environmentally friendly paper bags made from recycled paper. We believe there will be demand for this in the future.

6.6.4 Potential Customers and Needs

Companies are trying to break away from the use of plastic products and tackle environmental issues, an effort involving apparel companies interested in natural materials, persimmon farms, and government agencies and municipalities. The current trend in Japan is moving from charging for plastic bags to abolishing them, and the demand for durable Kakishibu coating will increase. Additionally, in Europe, where environmental awareness is high, Kakishibu, a natural and fermented material unique to Japan, is expected to become a big business opportunity from the perspective of the SDGs. Furthermore, the number of allergy sufferers due to chemical substances is increasing worldwide, and safe dyes are in demand. If Kakishibu is disseminated to the world through paper bags, the appeal of its former uses, such as house painting and clothing dyeing, will be revisited, and if the recognition of "kakishibu" increases, we believe that a new market will expand.

6.6.5 Profit Plan

The profitability plan is shown in Table 6.1.

In creating this business plan, we encountered the following challenges. First, the students struggled most with the cost-benefit analysis. They had to provide a basis for the pricing of products, sales, labor costs, and advertising expenses. It is very difficult to justify the price of a product without precedent. Also, it is hard to find convenient data even when searching the internet. The students had to rewrite the plan many times while receiving advice from me and financial experts. This is not something typically found in regular school textbooks. I believe it was a new learning experience. Second was the assumption about customers. Students only think of B to C (i.e., business to customer), but many businesses are B to B, and they need to learn about distribution. The more diverse the sales channels, the higher the likelihood of the business being successful. However, students had a hard time imagining what kind of sales channels exist. They proceeded while studying over time. The last challenge was about novelty. From the students' perspective, even if they think it is a new product, they may discover that there are many similar products when they conduct research. Therefore, creating products with new value becomes the most significant barrier.

We decided to present the completed business plan "Making Kakishibu a World Standard!" at a business contest. In Japan, business contests targeting students are held by universities, government agencies, and companies. If you win these contests, not only will you be featured in many media outlets, but you will also have

Table 6.1 Profit plan of Kakishibu paper bags

		After 1 year	After 5 years	Remarks
Sales		29.7 million yen	5809 million yen	Profit from selling Kakishibu paper bags
Cost of sales (purchase amount)		17.6 million yen	85.5 million yen	Kakishibu paper bags (raw materials)
Expenses	Personnel costs	9.6 million yen	12 million yen	President employees
	Rent	0.5 million yen	0.5 million yen	Office
	Advertising expenses	1.48 million yen	29 million yen	2% of sales
	Others	3.8 million yen	3.9 million yen	Communication and telephone expenses, utilities Consumable expenses, website maintenance costs, etc.
	Total	329 million yen	1309 million yen	
Profit		- 3.2 million yen	4500 million yen	

more opportunities to receive inquiries from interested companies and research institutions. This was expected to help promote the spread of Kakishibu.

The business contests we participated in and their final results are shown in Table 6.2.

The business plan proposed by the students using Kakishibu received high evaluations in many contests and garnered significant attention. For example, the students won the Minister of Environment Award at the Decarbonization Challenge Cup (Fig. 6.2). The judges evaluated the students' efforts to solve social problems using business methods from a new perspective, focusing on traditional resources in local regions in Japan. Another notable achievement was at the "9th U.S. Embassy-Keio SFC-TOMODACHI Entrepreneurship Seminar," which is an educational program established with the support of the Boeing Company aiming to nurture the next generation of leaders who will be responsible for Japan's recovery, regeneration, and development from the perspective of entrepreneurship, following the Great East Japan Earthquake. After a pre-seminar held in April 2019, 37 students from 18 high schools and universities nationwide who passed the preliminary screening gathered. After a joint seminar in August and the final selection in December, the students (Chikami, Ito, and Takaya) were selected as one of the three teams to undergo for training in the United States. As a result, they had the opportunity to present the potential of Kakishibu at Stanford University. Additionally, the result was recognized and featured in numerous media outlets, such as newspapers including the Kyoto Shimbun, the Yomiuri Shimbun, the Asahi Shimbun, and the Nikkei, as well

Table 6.2 List of business contests participated in and their results

Business contests	Date	Organizer	Features of the contest	Presenter	Final result	Evaluated points and comments
7th "Creativity, Infinity ∞ High School Business Plan Grand Prix"	January 12, 2020	Japan Finance Corporation	Over 3800 entries, a business contest for high school students	Chikami Ito Takaya	Special jury award	Proposed an idea for solving environmental problems as high school students using business methods
4th Japan University of Economics, High School Business Idea Contest	November 9, 2019	Japan University of Economics	Business ideas for achieving a sustainable society, themed on SDGs	Takaya	Best award	It should be commercialized immediately, and further development is expected in the future
SDGs Town Planning Idea Contest	November 9, 2019	Cabinet Office		Chikami	Second award	It involves the participation of the local community
Kansai University, Business Plan Competition KUBIC2019	October 5, 2019	Faculty of Commerce, Kansai University	Soliciting various venture business plans from high school students nationwide	Chikami Ito Takaya	Winner	Focusing on Japan's traditional resources and connecting them to beauty

(continued)

Table 6.2 (continued)

Business contests	Date	Organizer	Features of the contest	Presenter	Final result	Evaluated points and comments
Decarbonization Challenge Cup 2020	February 19, 2020	Japan Network for Climate Change Actions	Reporting regional activities related to global warming prevention, carried out by various entities nationwide, as a national movement for CO_2 reduction for the next generation	Chikami Ito Takaya	Minister of the environment award	No comments from the judges
5th National Youth Environmental Activity Presentation Meeting	February 8, 2020	Ministry of the Environment	Providing support to high school students who are conducting environmental activities towards the construction of a sustainable regional circular symbiotic society with opportunities for mutual exchange and practical presentations	Chikami Ito Takaya	President of the environmental restoration conservation organization award	The idea of not only using traditional Japanese technology to reduce plastic waste but also connecting it to regional revitalization was highly evaluated
Junior and Senior High School Future Creation Contest	February 22, 2020	The Yomiuri Shimbun	Soliciting ideas to achieve the Sustainable Development Goals (SDGs)	Chikami Ito Takaya	Best award	The approach of re-highlighting its effects from the perspective of environmental issues and broadcasting it to the world was highly evaluated

Business contests	Date	Organizer	Features of the contest	Presenter	Final result	Evaluated points and comments
9th U.S. Embassy-Keio SFC-TOMODACHI Entrepreneurship Seminar	April to December, 2019	U.S.-Japan Council— U.S. Embassy Tokyo Keio University	Aiming to nurture the next generation of leaders who will be responsible for Japan's reconstruction, regeneration, and development from the perspective of entrepreneurship	Chikami Ito Takaya	Dispatch to American research training (Stanford University)	The fact that they are attempting to link Japan's traditional resources to addressing environmental issues and sharing information globally
Challenge Future Award Business Design Discovery & Presentation 2019–2020	October 28, 2020	ICT Business Research Group	Seeking excellent business model designs, ideas, materials, etc.	Hidani Tahara	Minister of internal affairs and communications award	Society is demanding materials to replace plastic. The feasibility of this business is expected

Fig. 6.2 Decarbonization Challenge Cup 2020 (University of Tokyo), Winner of the Minister of Environment Award

as the Yomiuri Junior and Senior High School Shimbun (Fig. 6.3), and on TV stations like NHK and TV Tokyo. This led to many inquiries from companies interested in Kakishibu. As a result, the mayor of Kizugawa City, where Kizu High School is located, showed interest and requested to hear the story directly. When the students explained the potential of Kakishibu, the city expressed a willingness to fully support the endeavor. The mayor introduced the Agricultural Policy Division of Kizugawa City to the increased production of Kakishibu, the Tourism and Commerce Division to the introduction of Kakishibu paper bags as shopping bags, and the Environmental Division to dealing with the garbage problem, within a framework of industry-government-academia collaborations.

Furthermore, the students' activities caught the attention of the Ministry of the Environment, and they were invited to the Minister's office. They had a valuable opportunity to directly propose the potential of Kakishibu to the then Minister of the Environment, Shinjiro Koizumi. Before our visit, we asked the Minister for advance permission to demonstrate, in his office, the water resistance and durability of Kakishibu (Fig. 6.4). Although we were nervous that the newspaper might tear, which would be problematic, we successfully illustrated the appeal of Kakishibu. This experience was an important step in further emphasizing the significance of our project.

Fig. 6.3 Students featured on the front page of the Yomiuri Junior and Senior High School Shimbun with a prominent Japanese headline, meaning "The First Step Towards Changing the World"

6.7 Spreading Kakishibu to the World

After that, I transferred to two other high schools and have been continuing my efforts to explore the potential of using Kakishibu not only in paper bags but also as an alternative material for various plastic products. In this section, I would like to introduce my ongoing efforts to utilize it in cosmetics, agricultural mulch, in-flight

Fig. 6.4 Explaining the water resistance of the Kakishibu paper bag to the Minister of the Environment

etiquette bags, and food packaging paper. All of these are commonly used worldwide, and we expect them to be widely adopted not only in Japan but also globally.

6.7.1 Applying Kakishibu to Cosmetics

I formed a new team with students and participated in a business contest-style exploratory learning program, where high school students solved problems posed by companies and organizations. Among the eight companies proposing challenges, we decided on L'OCCITANE JAPON K.K.'s challenge: "Select two or more of the L'OCCITANE's six commitments, and propose a new sales strategy for L'OCCITANE that will increase sales and brand image by utilizing the selected themes." Among the six commitments (see L'OCCITANE's webpage https://loccitane.pearlbrandsonline.com/commitments/), we selected "supporting producers," "reducing waste," and "celebrating craftsmanship." After going through document screening, video screening, and online screening, we were selected as the best team by L'OCCITANE JAPON K.K. and received the Special Jury award for our business plan using Kakishibu for cosmetics, shown below.

6.7.1.1 Product Proposal: Hand Cream, Hair and Body Care Products, and Clean Hand Gel Containing Kakishibu

First, we proposed the development of a hand cream containing Kakishibu. Kakishibu has effective antibacterial and antiviral properties, and by incorporating it into L'OCCITANE's signature hand cream, a product with both high moisturizing power and the ability to inactivate pathogens and inhibit virus proliferation is born. With the heightened awareness of disinfection and hygiene due to COVID-19, a hand cream with both moisturizing and disinfecting functions is considered to meet market needs. In particular, by emphasizing that it can gently care for hands with natural ingredients while also providing hygienic measures, it can be differentiated for those suffering from rough hands due to alcohol disinfection.

Next, we proposed the development of hair care and body care products that utilize the deodorizing function of Kakishibu. By incorporating Kakishibu into L'OCCITANE's popular hair care series "Repairing," we can launch new products with high deodorizing effects. The phenol hydroxyl groups contained in Kakishibu chemically bind with odor components, transforming them into new odorless components, and exhibit excellent deodorizing effects. They also inhibit the growth of bacteria on the skin that cause body odor, so they can also help suppress odors caused by sweat and sebum. Beyond hair care products, by applying Kakishibu to body washes and body lotions, we can launch a range of products that can prevent body odor while caring for hair and body. This allows us to attract new customers from a wide range of generations, including men who are concerned about body odor, not just women.

Finally, we proposed the development of a clean hand gel that maximizes the antibacterial and antiviral effects of Kakishibu to maintain hand hygiene. The antibacterial effect of Kakishibu is more persistent than alcohol, and the effect lasts for a long time even after disinfection, making it an excellent product from the perspective of infection prevention. The disinfection market is experiencing rapid growth due to the impact of COVID-19. In response to this market expansion, we aim to establish a new source of revenue by offering products made with natural ingredients. In addition, to further strengthen L'OCCITANE's environmental conservation efforts, we propose the introduction of paper packaging using Kakishibu. L'OCCITANE is promoting plastic reduction through the use of 100% recycled plastic. However, by substituting paper containers with the waterproof and antibacterial properties of Kakishibu, we can achieve a high-quality package design while reducing environmental impact, which also leads to an improvement in brand image.

6.7.1.2 Market Strategy and Sales Channels

This product will be primarily sold through L'OCCITANE's existing sales channels such as directly managed stores, e-commerce, and department stores, while also expanding into the amenity market of luxury hotels. L'OCCITANE's amenities are adopted at five-star hotels, and PRbenefits can be expected through the brand experience of guests. By introducing products containing Kakishibu as amenities, it is possible to appeal to consumers with a preference for natural products creating an effect that leads to future purchases.

6.7.1.3 Procurement Plan and Cost Estimation

The raw material for Kakishibu can be supplied from Mimasu Kashichi Shouten Co., Ltd. in the Yamashiro region of Kyoto Prefecture. The company has stable supply capacity in carrying out the production, juicing, fermentation, and sales of astringent persimmons in a consistent manner. To demonstrate the functionality of Kakishibu, it is necessary to have an antibacterial effect of 0.1% or more and a deodorizing effect of 0.5% or more. If 1% is mixed into hand cream (30 mL), the cost is estimated to be about 5 yen per 30 mL hand cream. In addition, to ensure a stable supply, we will promote the cultivation of astringent persimmons in hilly and mountainous areas, considering environmental conservation effects such as the use of abandoned agricultural land and the prevention of sediment disasters. This will build a sustainable production system and contribute to the realization of fair trade.

6.7.1.4 Sustainable Business Model

This proposal was based on the philosophy of L'OCCITANE, aiming to fuse traditional Japanese technology and environmental considerations. By disseminating Kakishibu technology that has been passed down since the Heian period to the world, we aim to use traditional technology to contribute to environmental protection by strengthening plastic reduction strategies. Also, realizing fair trade with persimmon farmers will support the producers. The era ahead requires a new set of values that strike a balance between convenience and environmental considerations. By using Kakishibu, a traditional natural material from Japan, we can merge French and Japanese cultures and are confident that L'OCCITANE can provide new market value. We received positive feedback indicating that our proposal aligns with L'OCCITANE's commitments, and they expressed a desire to work together as a step toward a more beautiful future.

6.7.2 Applying Kakishibu to Agricultural Mulch

We are also working on the development of paper mulch coated with Kakishibu as an alternative to the vinyl mulch used in agriculture. Vinyl mulch is widely used mainly in open-field vegetable cultivation for its effects such as weed control, increasing soil temperature, soil moisture retention, and prevention of fertilizer runoff. However, its disposal not only imposes an environmental burden but also presents challenges such as rising collection costs and heavy labor involved in processing (Ministry of Agriculture, Forestry and Fisheries, 2024). Currently, there are companies in Japan that manufacture and sell paper mulch; however, it is susceptible to water damage and has durability issues. Therefore, we aim to develop a product that coats paper mulch with Kakishibu, improving its water repellency and durability, and ultimately decomposes by being absorbed into the soil. The appeal of this product is that it is made entirely of natural materials and can be returned to the soil after use, reducing the burden of agricultural work to remove them. In addition, the antibacterial, insect-repellent, and mold prevention effects of Kakishibu can also reduce the use of pesticides and fungicides. We hope that the Kakishibu paper mulch we propose will become a popular sustainable agricultural material, leading to changes in agricultural practices worldwide.

6.7.3 Applying Kakishibu to Papers for Food and Drink

We are planning to apply Kakishibu to papers for food and drink such as food packages. Oil and water resistance are the most important functions in food packaging paper, and per- and polyfluoroalkyl substances (PFAS) were used to enhance the functionality of ordinary paper, which has low durability against water, oil, and heat (Ministry of the Environment, 2025). However, it has now been revealed that these coating agents containing PFAS have cumulative, harmful properties (Ministry of the Environment, 2025). As a consequence of considering impacts on the human body and the natural environment, the use of PFAS is beginning to be regulated in Japan (Ministry of the Environment, 2025).

In recent years, there has been a gradual shift from plastic to paper in Japan. Examples include straws, cups, plates, and food containers. In the face of further environmental burdens and health risks associated with PFAS processing, we believe that enhancing the water resistance and durability of paper with natural materials like Kakishibu can open up a new market.

6.8 Things Only High School Students Can Do

When I was assigned to Kizu High School and began working with Kakishibu, not many people were interested in it, except for some local stakeholders. Despite Kakishibu having a history of more than 1000 years, it has been gradually forgotten, especially by the younger generation. This may have been due to the perception that "Kakishibu is no longer a necessary resource." However, through efforts by high school students disseminating information with a new perspective, we were able to attract the attention of many people.

In the process of developing a business plan for Kakishibu, the high school students showed changes such as "I started to pay attention to social issues that I thought had nothing to do with me" and "I realized that everything is inter-related." The students experienced significant growth, both personally and professionally, through their activities.

For example, a student who was the activity leader at the time chose Kizu High School because it was close to home and she wanted to relax at home after school. However, through her involvement in Kakishibu projects, she gradually developed a proactive attitude. At one point, she sent a letter to the headquarters of a global company, proposing, "Why don't you change the bags you use in your store to Kakishibu paper bags?" It was a long shot, but unexpectedly, they received a positive response, and it was decided to introduce them on a trial basis in stores within the prefecture. "I can't believe such a big company would respond to an initiative by a rural high school," she noted, thrilled.

Furthermore, she reflects, "Through the activities, I learned that there are many wonderful adults. People with light footwork and full of curiosity took an interest in our activities and supported us. I came to admire those who became 'good adult role models' for us."

There were only three members involved in the activity at that time, but their horizons expanded, and all of them used their high school activity achievements to advance to universities through holistic admissions, eventually being admitted to programs in the faculties of commerce, agriculture, and education. I have always thought that I want to shorten the distance between high school and society. By connecting with society, students become more independent and start to think for themselves. Given the right educational environment, students have the power to grow on their own. As we have seen so far, I think we can summarize the "things only high school students can do" into the following three capacities.

First, unlike adults, high school students can freely tackle real-world challenges with their open minds, whereas companies are required to pursue profits, and there may be limitations on activities due to factors like position, organization, and time.

Second, I believe they can afford to fail. If you treat it like a job, unless it is volunteer work, there is a risk of losing your job or income if you fail. But if you are a student, you can see failure as part of the learning process. I believe failure leads to growth.

Finally, society often values and supports students' efforts. This is because I believe that students' challenges bring new ideas that are not limited by existing frameworks and can lead to fresh solutions to regional issues. Additionally, the sincere attitude of young people, who carry the future, towards socio-technical issues fosters empathy and support, encouraging adults to connect with the next generation. Moreover, collaboration between education and the local community is not just a one-off activity but also holds the potential for sustained use of local resources and industrial growth. So far, by addressing socio-technical issues, we have gained cooperation from many companies and research institutions.

Certainly, there remain numerous challenges within the Kakishibu business, and an abundance of obstacles to surmount. Nonetheless, this is precisely why I anticipate that the students will approach the future of the global environment with confidence and resilience, viewing it from a distinctive student perspective.

6.9 Conclusion

As we have observed so far, we have consistently conducted activities that could be called "socio-technical experiments with Kakishibu," with students leading the efforts. An overview of Kakishibu stakeholders involved in the experiments is shown in Fig. 6.5.

While tackling the microplastic problem, the students focused on the functionality of Kakishibu, a traditional resource. They advanced the project while receiving advice from Mimasu and Matsuo, experts in Kakishibu research and practices, and were able to gain the cooperation of corporate companies and universities by disseminating information. I believe that these connections with companies and universities would not have been established if the high school students had not taken the first step in their activities. Furthermore, the circle grew as students challenged themselves in business contests. Ongoing information sharing through the media and the internet paid off, allowing us to attract the interest of the Ministry of the Environment and global companies.

The following three characteristics can be identified in such student-led innovation with business.

Firstly, students can take on the role of risk-takers. The Kakishibu experiment also started as an initiative with little precedent, without budget or facilities. Initially, it was a series of failures, but through repeated attempts, we achieved promising results.

Secondly, students can attract more attention. Until now, the local companies and related organizations that have been working on the popularization of Kakishibu have not been able to attract enough interest. Also, I believe that volunteer activities

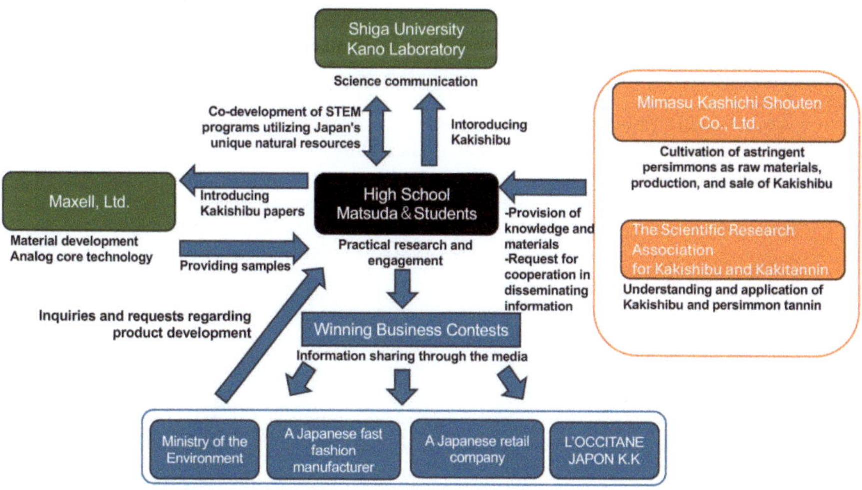

Fig. 6.5 Students-centered overview of stakeholders involved in the socio-technical experiments with Kakishibu

ended because these were temporary efforts. However, when the students proposed business-related initiatives, a lot of interest was generated, and the initiative showed significant growth.

Lastly, the potential of students is infinite. Due to the activities of the students, global companies and government agencies have started to move. I have seen many times where the students' seemingly reckless approach has moved society. I am convinced that the enthusiasm of the students has the power to move the hearts of adults.

Acknowledgments I feel a certain destiny in discovering research on Kakishibu. When I was sent as a trainee to Kyoto University for half a year in 2013, I was cared for by the late Professor Hino Motosugi at the Graduate School of Life and Environmental Sciences at Kyoto Prefecture University, who was an expert in fruit trees and did collaborative research with colleagues in Kyoto University. Professor Motosugi emphasized the importance of Kakishibu research in the Minamiyamashiro region to pass it on to future generations, and worked hard to establish the "The Scientific Research Association for Kakishibu and Kakitannin" with Professor Emeritus Matsuo and others. However, I was shocked to learn from his wife's New Year's card that he had died in a hit-and-run accident at the young age of 54. At that time, I did not know that Professor Motosugi was researching Kakishibu. Later, when I transferred to Kizu High School in 2018 and began working on Kakishibu projects with students, I saw Professor Motosugi's name and felt as if I were being guided by him. Through my work with Kakishibu, I was blessed with many connections to students and gained valuable experience. Now, many researchers and companies recognize the potential hidden in Kakishibu. I believe this is part of the charm of Kakishibu, nurtured by Japanese tradition and culture.

References

Association of Marine Leisure Promotion (2023) *Fuiripin no Borakai to no jirei* (case study of Boracay Island in The Philippines). https://amp.okinawa/tourism10/

Japan Science and Technology Agency. (2009). *Heisei 20 nendo kotogakko rikakyoin jittai chosa* (survey on the actual conditions of high school science teachers in 2008).

Japan Science and Technology Agency. (2021). *Supa Saiensu Haisukuru jisshi yoko* (super science high school implementation guidelines).

Kobayashi, S., & Higano, Y. (2000). A study on artificial forests: An estimate for the public interest and a proposal for management policy. *Studies in Regional Science, 31*(3), 237–250.

Ministry of Agriculture, Forestry and Fisheries, Crop Production Bureau, Horticultural Crop Division. (2024). *Engei bun'ya kara haishutsusareru purasuchikku o meguru josei* (Situation regarding plastic emitted from the horticultural sector). https://www.maff.go.jp/j/seisan/ryutu/engei/sisetsu/attach/pdf/used_plastic-4.pdf

Ministry of Education, Culture, Sports, Science and Technology. (1978). *Koto gakko gakushu shidoyoryo kaisetsu* (High school curriculum guidelines).

Ministry of Education, Culture, Sports, Science and Technology. (2009). *Koto gakko gakushu shidoyoryo* (High school curriculum guidelines).

Ministry of Education, Culture, Sports, Science and Technology. (2018). Koto gakko gakushu shidoyoryo kaisetsu: Sogoteki na tankyu no jikan (Explanation of high school curriculum guidelines: The period of integrated study).

Ministry of Education, Culture, Sports, Science and Technology. (2024). *Kotogakko gakkabetsu seito su gakkosu* (Number of students and schools by high school department). https://www.mext.go.jp/a_menu/shotou/shinkou/genjyo/021201.htm
Ministry of the Environment. (2025). *PFAS handbook.*
Mitsubishi Research Institute. (2019). *Seibunkaisei purasuchikku no kadai to shorai tenbo* (Challenges and future prospects for biodegradable plastics). https://www.mri.co.jp/knowledge/column/20190408.html.
Munesue, Y., & Masui, T. (2012). Long-term evaluation of the impact of bio-based plastic diffusion on global food insecurity. *Environmental Science, 25*(3), 167–183.
Nakamura, T., Matsuo, T., Okamoto, S., Tomokane, I., Mori, T., Sato, K., & Otsuki, K. (2004). Bioantimutagenicity of heirloom vegetables in Kyoto. *Environmental Mutagen Research, 26*(3), 259–264.
Nakamura, Y., Nakayama, Y., Ando, H., Tanaka, A., Matsuo, T., Okamoto, S., Upham, B. L., Chang, C. C., Trosko, J. E., Park, E. Y., & Sato, K. (2008). 3-Methylthiopropionic acid ethyl ester, isolated from Katsura-Uri (Japanese pickling melon, *Cucumis melo* var. *conomon*), enhanced differentiation in human colon cancer cells. *Journal of Agricultural and Food Chemistry, 56*(9), 2977–2984.
Namimatsu. (2020). Kyoto food culture and intangible cultural heritage "Washoku": The relationship between the history of Kyoto cuisine and the Japanese diet. *The Bulletin of the Institute for Japanese Culture, Kyoto Sangyo University, 25*, 200–236.
Nihon Shimbun Kyokai (The Japan Newspaper Publishers & Editors Association). (2024a). *Shinbun no kobetsu haitatsuritsu* (Newspaper door-to-door delivery rate). https://www.pressnet.or.jp/data/circulation/circulation03.php
Nihon Shimbun Kyokai (The Japan Newspaper Publishers & Editors Association). (2024b). *Shinbun no hakko busu to setaisu no suii* (Changes in newspaper circulation and number of households). https://www.pressnet.or.jp/data/circulation/circulation01.php
Nishikawa, Y., & Hamaguchi, M. (2018). Shushi wo meguru shimin soshiki, nomin soshiki no kokusaiteki jokyo ni kansuru kosatsu – Shokuryo oyobi nogyo no tame no shokubutsu iden shigen ni kansuru kokusai joyaku dai 7 kai teiyaku koku kaigi sanka wo tsujite – (Reflections on the international situation of citizens' and farmers' organizations for seeds: Through participation in the seventh meeting of the conference of the parties to the international treaty on plant genetic resources for food and agriculture). *Kurumedaigaku Keizai Shakai Kenkyu (Economic and Social Research, Kurume University), 58*(3–4), 101–125.
Sakata, H. (2010). Kosaku hoki ni taisuru shika inoshishi ni yoru higai no eikyo (The impact of damage caused by deer and wild boar on abandoned cultivation). *Hyogo Wildlife Monograph, 2*, 36–38.
Takashima, S. (2003). *Saijiki: Kyo no dento yasai to shun yasai (Kyoto's traditional vegetables and seasonal vegetables).* Tombow Publishing.
United Nations. (2024). *World population prospects 2024: Summary of results.*
Yamasaki, M., & Watanabe, T. (2024). Development and impact of environmental policy in EU, UK, and France: Focusing on plastics-related policies and food loss and waste reduction policies. *Senshu University's Bulletin of the Research Institute of Commerce, 55*(3), 1–55.
Yoshikawa, S. (2015). Deforestation and human pressure on the Brazilian Amazon rainforest. *Journal of the Remote Sensing Society of Japan, 35*(4), 260–261.

Toshihiko Matsuda is a teacher at Ikubunkan Junior High School, Ikubunkan High School, and Ikubunkan Global High School, all part of the Ikubunkan Institute of Education. He leads classes in information and technology and conducts inquiry seminars. After earning a bachelor's degree in biological resources and a teaching license from Hiroshima Prefectural University, he was hired as a high school agriculture teacher by the Kyoto Prefectural Board of Education. At Katsura High

School, he started a program with students focused on a seed bank for kyo-yasai, traditional vegetables in Kyoto, and served as a committee member on the Kyoto Vegetable Functional Utilization Promotion Liaison Council, established by Kyoto Prefecture. At Kizu High School, he launched a program with students to promote Kakishibu, was recognized for his efforts—such as guiding business contests—and received the Kyoto Prefecture School Excellent Teacher Award.

Chapter 7
Industrial Challenge: How to Work with Traditional Knowledge for Commercialization

Toshiharu Goto, Yuka Yokoyama, and Shota Taniguchi

Abstract In this chapter, we present examples of applying Eastern techniques and ideas that can be called indigenous knowledge to actual manufacturing sites, discussing the difficulties and challenges involved in the development process of new patents.

In developing a waterproof coating using Kakishibu, we discuss the positioning of development of paints and coating from the perspective of patents owned by Maxell, Ltd. Based on this, we show how the company recognized the value of traditional techniques. We also discuss the problems associated with the commercialization of traditional techniques when viewed as modern practices, citing other examples found in the literature. Additionally, we explain the management approach based on Eastern philosophy, including Buddhism, that was used to promote the research in this case.

Furthermore, we describe the technical review undertaken, including forming the development concept, experimental method, experimental data results, and evaluation methods. This allows us to discuss the possibility of industrializing traditional techniques, which are time-consuming and labor-intensive. We try to show that the fusion of traditional techniques and modern science is an effective way to promote the commercialization of traditional techniques.

Finally, we discuss the potential applications of traditional techniques in modern industry and society.

Keywords Industrialization · Patent · Water repellent · Internship · Management · Japanese Umbrellas · Peril oil · Lecithin · Ultraviolet · Minakata Kumagusu · Mandala · Buddhism

T. Goto (✉) · Y. Yokoyama · S. Taniguchi
New Business Management Division, Maxell, Ltd., Kyoto, Japan
e-mail: toshiharu.goto.te@maxell.co.jp

K. Kano et al. (eds.), *Traditional Knowledge for Green Innovation with Japanese Persimmon Tannin, Kakishibu*, https://doi.org/10.1007/978-981-95-7805-4_7

7.1 Introduction

Goto, the first author, would first like to discuss why Goto became interested in Kakishibu, a traditional technique used in Japan for many years on paper, wood, and clothes. Goto moved to Kyoto and joined Maxell, Ltd., which was established in Kyoto in 1960 to produce dry batteries, because Goto was asked to conduct product development and research that would lead to future technology. Goto was also considering whether to develop environmentally friendly products using his background in polymer science and coating technology. In this context, Goto was introduced to Kakishibu by Professor Sadao Harada, an Associate Professor of economics at Doshisha University, who was giving a lecture in Shiga Prefecture, known for its high environmental awareness. At that time, Goto felt a connection with Professor Harada, and he may have sensed Goto's freedom to explore new ideas. A few months later, Professor Harada introduced Goto to the efforts of high school students and their teacher, Toshihiko Matsuda (then at Kizu High School), to utilize Kakishibu. We soon visited Mr. Matsuda. We were excited about the traditional technique of improving the water resistance of paper with Kakishibu, which was a fresh concept for us, and our desire to explore product development using Kakishibu grew.

In the section, we would like to explore the problems we faced during our development efforts at Maxwell in the four years after these meetings. We will also discuss the difficulties involved in promoting the industrialization of traditional techniques in a broader economic sense.

7.2 Technical Value Evaluation Within the Company

As mentioned in Chap. 6, what was being considered at Kizu High School, where Mr. Matsuda was affiliated at the time, was the application of Kakishibu to newspapers. Newspapers are inherently weakened by water, so the effect of Kakishibu could be more beneficial. In the case of newspapers, high mechanical strength is not required as they are not expected to be folded complexly or used repeatedly. Therefore, paper could be made by connecting short fibers with a binder instead. Thus, the newspapers have lower strength than usual papers, but the water-resistant effects of Kakishibu became clear.

Existing literature explores cases where Japanese paper was strengthened with Kakishibu. In a study conducted by Oda (2014) that engages with the knowledge of kamiko, paper clothes and fashion works were created using washi, a traditional Japanese handmade paper coated with Kakishibu. In this study, *washi* that had been kneaded and coated with Kakishibu twice with a brush was used, a similar process to that in the Edo Period (1603–1868). It was shown that the mechanical properties clearly improved, and water resistance slightly improved.

Inspired by the traditional works with *washi*, we decided to use kraft paper, which is usually used for packaging, for our study because *washi* is expensive and not so popular compared to Kraft paper. Kraft paper is designed to be folded and used, and therefore, it would be a problem if it became damaged when folded. As such, long fibers obtained from coniferous trees are used as raw materials. Such paper does not repel water, but the mechanical properties of the paper itself are high. Also, we used kraft paper as it is, not kneaded paper (e.g., paper clothing). We conducted tests that, even when using relatively thin kraft paper of 50–70 g/m^2, the effect of Kakishibu was not clearly seen in tests such as tensile tests of wet paper. Also, at this stage, we could not find a clear difference in terms of water repellency. Maxell Ltd. Corporation's research and product development policy prioritized the utilization of its proprietary technologies or patents to emphasize distinctive characteristics. Consequently, the mere application of Kakishibu was inadequate for establishing a viable research framework because the corporation has not had enough know-how relating to Kakishibu. To explore the potential of research and development utilizing Maxell's technology, we conducted a survey to classify and count the patents applied for by Maxell, Ltd. after January 1, 2004, by field. In this process, we used a classification system called F-term, compiled by the Japan Patent Office. The F-term is a classification system introduced in 1987, coexisting with the International Patent Classification and FI used by the Japan Patent Office, and is suitable for computer-based searches. Using the theme codes in this classification system, we could survey the patents applied for by Maxell, Ltd. Figure 7.1 shows the top 30 theme codes obtained from this classification. Patents related to televisions and imaging transferred from Hitachi Ltd., and battery-related patents, occupy the top positions. Other patents related to optical components and recording media, which were once our main products, are also scattered.

At first glance, these technologies seem unrelated to the development of water-resistant paper. However, recording media and batteries are products manufactured by applying paint to a film. Paint formulation and coating technology will be key considerations when developing water-resistant paper using Kakishibu.

Therefore, we classified Maxell's technologies from this perspective. Specifically, we classified patents that include the keyword "paint" in the specification using the F-term. The results are shown in Fig. 7.2. Many patents for magnetic recording media are identified when searched for from the perspective of paint. Patents related to precision coating and paint formulation/dispersion technology can also be picked up. Moreover, there are more than 160 patents even in the classification of paint. Therefore, we considered proceeding with the study in a direction that would bring out unique technical features with the keywords "paint formulation" and "coating." We also proposed aiming for a technology with high biodegradability and other environmental compatibilities by using natural materials as much as possible. However, the company did not accept the idea of generalizing Maxell's paint and coating technology and applying it to the development of paint using Kakishibu and thus failed to establish it as a formal project with a dedicated budget personnel. This example may be an unfortunate example of a company that is unable to see its

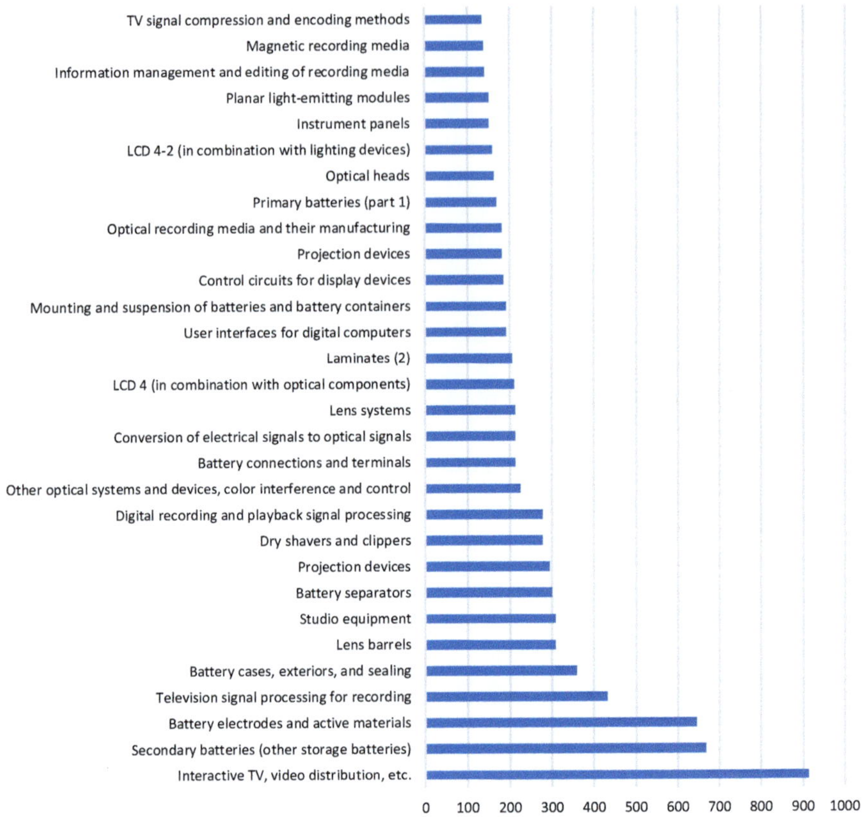

Fig. 7.1 The patents applied by Maxell, Ltd.

technology as the basis for expanding to broader applications, thereby hindering possibilities for innovation.

7.3 Empathy Within the Company

7.3.1 Economic Effects and Investment Direction

In this section, we discuss whether we could get support within the company to promote research on traditional technologies. According to Wang and Takahashi (2020), while an increase in research and development costs contributes to innovation, its efficiency may have been declining in recent years. This trend is reported in various fields worldwide, not limited to Japanese companies (Gordon, 2016; Bloom et al., 2020). Traditionally, companies promote research and development investments following Joseph Schumpeter's idea that innovation is the source of

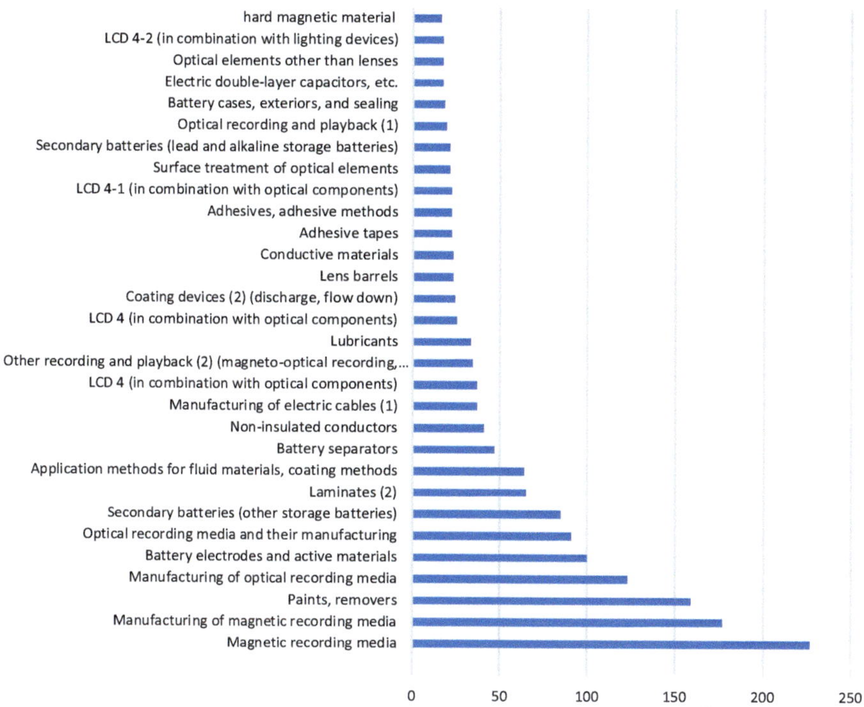

Fig. 7.2 The patents applied by Maxell, Ltd. that include the keyword "paint"

economic growth. However, if companies believe that investment efficiency is declining, they narrow the direction of development investment, and it is reported that Japanese companies have been promoting concentration and selection since the 2000s (Nakanishi, 2022; Orihata, 2006). On the other hand, in the report by Nakanishi (2022), it is pointed out that these concentrations and selections are not necessarily leading to an improvement in the profit margin.

With respect to environmental development investment, according to a report by Edamura et al. (2017), this investment in Japanese manufacturing has been increasing since 2000, with the main investors being manufacturers of products with environmentally harmful images, such as the petroleum industry, automobile industry, and cement and cement product industry. On the other hand, in manufacturers of electronics and electrical equipment like Maxell, Ltd., the proportion of research investment in the environmental field is not high. Meanwhile, data analyses revealed that environmental investment enhances brand assets and corporate value more than social activity investment or advertising investment.

In terms of concentration and selection of research investment, Maxell, Ltd. clearly focuses on investment in battery-related fields. The fields of research using Kakishibu are expected to be food, agriculture, and apparel. These fields are not in the electrical and electronic sectors that Maxell, Ltd. is primarily engaged in, making it difficult to create a story that clearly builds on Maxell's existing business interests. As mentioned earlier, in the electrical and electronic industry, the proportion of research investment in the environmental field is low. On paper, it would seem that Kakishibu would be difficult to justify investing in. Also, when looking at Maxell's approach to environmental research investment, there is no discussion about the efficiency of advertising expenses or social activity investment to determine their investment allocation. Therefore, the reality is that there is no concept of conducting in-house research and development investment as advertising expenses or social activity expenses. This research project, which aims to apply the traditional technology of Kakishibu to environmental studies, was expected to be viewed with skepticism regarding its economic value within the company.

7.3.2 Comparison with Petroleum Products: Skepticism About Economic Value

Here, we will discuss the positioning of traditional industries and natural substances against petrochemical products, while touching on traditional industries in modern society. In Japan, traditional industries and technologies have been maintained by diversifying uses, arranging designs of traditional products in a modern style, or applying technologies to objects that are easy for modern people to use.

For instance, in Sanjo, Niigata Prefecture, known for its production of blades, the manufacturing of knives and forks began in the Meiji era, and it is still recognized as a city of blades; along with Tsubame City, this area is collectively known as Tsubame-Sanjo (Goto & Takashima, 2021). The cutlery produced here has even been used at Nobel Prize award ceremony banquets. In other examples, Sony has utilized the fire chopstick technology from Himeji City, Hyogo Prefecture, for microphone checks, and Hitachi Metals has revived and preserved the traditional "*tatara*," a traditional Japanese furnace used for smelting iron and steel, used in sword making in Shimane Prefecture, demonstrating efforts to utilize technological heritage (Shimono, 2003). In Fukuyama City, Hiroshima Prefecture, the manufacturing techniques and knowledge of Bingo Kasuri, which has been ongoing for over 400 years, is being utilized in denim manufacturing, holding a high domestic market share (Administrative Evaluation Bureau, Ministry of Internal Affairs and Communications, 2022). On the other hand, there are many instances where techniques used in everyday manufacturing are being lost due to changes in lifestyle habits. One such example is the technology using Kakishibu, with many Kakishibu manufacturers being forced out of business (Imai, 2003). To prevent the loss of such culture and technology, it is necessary to increase the economic value of traditional industries. A hint for this could be found in the study of the economic value of traditional crafts (Yokogawa & Nakano, 2016). According to this study, the element of storytelling is a factor that consumers prioritize over price. In terms of this storytelling aspect, the paint currently under development is considered perfect. High school students in Kyoto focused on the local material, Kakishibu, and this inspired engineers at Maxell, Ltd. to start thinking about actively utilizing this tannin industrially.

Maxell's headquarters is located on the only place where oil trading took place in the Edo period, and it is home to the only shrine in Japan related to oil. Noticing this, Mr. Matsuda of Kizu High School proposed the use of oil.

In order to utilize this idea, university interns were involved in the development of a paint made from oil and Kakishibu, resulting in a technology with a dramatic backstory. Smelting iron and steel were used in sword making in Shimane Prefecture, to demonstrate using technological heritage (Shimono, 2003). And in Fukuyama City, Hiroshima Prefecture, the manufacturing techniques and knowledge of Bingo Kasuri, which have been ongoing for over 400 years, are being utilized in denim manufacturing, possessing a high domestic market share (Administrative Evaluation Bureau, Ministry of Internal Affairs and Communications, 2022).

On the other hand, there are many instances where techniques used in everyday manufacturing are being lost due to changes in lifestyle habits. One such example is the technology using Kakishibu, with many Kakishibu manufacturers having been forced out of business (Imai, 2003). To prevent the loss of such cultural knowledge and technology, it is necessary to increase the economic value of this traditional industry. Useful insights for this can be found in the study of the economic value of traditional crafts (Yokogawa & Nakano, 2016). According to this study, storytelling is a factor that consumers prioritize over price.

In terms of storytelling, the paint currently under development by high school students in Kyoto is a promising opportunity. These students focus on the local material, Kakishibu, and this inspired engineers at Maxell, Ltd. to start thinking about actively utilizing this tannin for industrial purposes. It was by coincidence, but Maxell's headquarters are located in the only place where oil trading took place in the Edo period, and it is home to the only shrine in Japan related to oil. Noticing this, Mr. Matsuda of Kizu High School proposed the use of oil together with Kakishibu. In order to utilize this idea, university interns were involved in the development of a paint made from oil and Kakishibu, resulting in a technology with a dramatic story. The details of the internship will be provided later in Sects. 7.4 and 7.5.

However, a fundamental problem is that, with respect to particular valued properties, natural materials tend to fall short when compared to products made from petrochemicals. For example, the structure of natural fibers depends on the raw material, so there is limited room for improvement. On the other hand, synthetic fibers synthesized by chemical reactions can be further improved for higher performance and functionality, offering greater control over physical properties (Miyazawa, 2018). There are also attempts to make films from natural substances such as cellulose and chitosan (Hosokawa & Nishiyama, 1990), but it is difficult to directly replace the synthetic polymers currently in widespread use due to issues with water absorption and differences in molding methods.

Shifting the perspective slightly, polymers synthesized from bio-based substances, such as polylactic acid, which are expected to replace petrochemical products, are also becoming more widespread. However, even these new natural-derived synthetic polymers have difficulty replacing existing polymers in terms of characteristics (Shekhar & Mondal, 2024). Moreover, these materials are not technologies that use natural substances as they are, so there are many unknowns about the impact on ecosystems, such as the rate of biodegradation and the impacts of byproducts when these materials are discarded or released into the environment (Ali et al., 2023). Thus, even polymers synthesized from natural raw materials are inferior in terms of functionality and are more expensive than general-purpose polymers. Therefore, it is quite difficult to clearly define the economic value of products that are processed from natural materials. As modern society is accustomed to using cheap, high-performance synthetic compounds, finding value in materials made from natural materials can be a challenging proposition.

Facing this difficult situation is the challenge of our Kakishibu project, and it is undeniable that there are challenging hurdles to clear in obtaining full support from profit-oriented companies. However, processed products of washi, a type of paper that has been used for hundreds of years in textile products, should clearly be suitable based on their track record. Although it is a technology that is being lost due to economic rationality, if we can revive it with modern technology, we believe that it makes a valuable contribution to the future of humanity and the environment.

In Japan, the fact that the paper has been replaced by Western paper, which has more limited uses than *washi*, also seems to be an obstacle. *Washi* has longer fibers and is significantly stronger than Western paper (Yamauchi & Usami, 2005), which is why it has been used for clothing in Japan. However, very few people today can readily imagine using "paper" for clothing. Along with the technical issue of producing good quality products at low cost, there is a need to change the user perception of the material. Popularizing this traditional technology in society not only requires efforts of engineering but also social initiatives including education and politics.

7.4 Breaking Through Difficult Situations with Limited Funds and Personnel: Leveraging Internships

Goto believed that developing paper using Kakishibu could become a highly promising area of research and development, as it would enable the utilization of Maxell's patented precision coating technology and contribute to solving environmental issues, even under less supportive environment in the company. Therefore, Goto consulted with various acquaintances within the company about the situation where we could not proceed with the research because of not allocating budget and personnel. Not all opinions were negative about our Kakishibu project. Goto happened to have a causal relationship with the head of the CSR department at the time, and when Goto talked about this issue, the head suggested that we can accept an intern and give them this theme. At that time, our CSR department was accepting interns from universities and compiling opinions from interns about our CSR based on Maxell's public information and external information. However, there was no precedent for accepting interns as proactive researchers in the field of research and development, so the following points were raised as issues:

1. Ensuring safety.
2. Insurance issues.
3. Handling of intellectual property (IP).

Insurance (2) is an administrative matter, so this will not be examined here. There are several safety (1) considerations related to traditional technology. First, it is thought that traditional technologies use few purified chemical substances or chemical substances obtained by synthetic reactions. As a result, these technologies tend to involve materials with relatively mild physical and chemical properties. In particular, since the emphasis is on using natural materials as they are, plant extracts and foods must be safe to handle for an intern. Experimentation would involve mixing such materials, drying them, and using water to confirm water resistance; there are no dangerous substances involved.

On the other hand, IP (3) is related to Japanese law concerning the attribution of inventions (Patent Office of Japan, 2003). In Japan, the right to patent an invention made in the course of employment primarily belongs to the inventor (employee). However, a company can stipulate in a contract or employment rules that this right is attributed to the company. In such cases, the company is obligated to pay a "reasonable compensation" to the employee.

However, interns are not in an employment relationship. Therefore, if an intern invents something during their internship, the rights to that invention initially belong to the intern. In other words, if an intern comes up with an idea that leads to an invention, and the company wants to apply for a patent for this invention or has already applied, the intern has the right to refuse to transfer the rights to the company. Even if they agree to the transfer, they have the legal right to negotiate for compensation for the transfer. This point was raised as a significant issue when entrusting research to interns. We consulted with the relevant departments within the company and ultimately developed a scheme that would not be disadvantageous to either the interns or Maxell, Ltd.

As a result, we were able to proceed with research and development by accepting interns. However, since there is no dedicated employee for each intern, we had limited manpower. Therefore, Goto, one of the authors, first thought that it would be better for interns to come intermittently over a long period of time rather than coming intensively for a short period of time. When considering new materials, the process cannot be advanced intensively because it requires time from the initial proposal to material procurement. It is also difficult to maintain a situation where new ideas arise in a short period of time. As a result of consulting with the Career Center at Kobe University, which has been cooperating with us, we were able to implement a program that lasts for one year at a frequency of once a week.

7.5 Research and Development by Interns

Having finally established the internship environment, we promptly commenced the development of water-repellent paper utilizing Kakishibu. Although our team comprised only an intern, our development funds remained extremely limited. However, since producing Kakishibu-based water-repellent paper on a large scale necessitated cost-effectiveness, the constrained funds ultimately proved advantageous. Section 7.5.1 describes the research and development process and outcomes led by intern Yokoyama. Section 7.5.2 outlines the research and development efforts and results led by Taniguchi, who joined after Yokoyama and further elaborated upon Yokoyama's work.

7.5.1 Development of Water-Resistant Paper Using Traditional Japanese Umbrella Craftsmanship

Yokoyama, one of the authors, first joined the Kakishibu project in Maxell Ltd. The first research and development supervised by Goto involved using natural materials together with Kakishibu to impart water resistance to Kraft paper. This is because we should avoid using synthesized chemicals to maintain Kakishibu's environmentally friendly qualities. There are several indicators for water resistance, but in this context, we define water resistance as the paper itself not absorbing water. One possible method to improve the water resistance of paper is the application of a water-resistant coating. Attempts to impart water resistance to paper by applying materials have been made for a long time, and the technique used in the production of Japanese umbrellas is one such example. In the case of Japanese umbrellas, water resistance is achieved by applying drying oils such as perilla oil to the paper (Imai, 2003). Therefore, we hypothesized that applying perilla oil, a drying oil, to Kraft paper could impart water resistance to the paper.

For this study, we prepared Kraft paper (manufactured by Hokuetsu Corporation) with a starting weight of approximately 50 g/m^2 and a size of 15 cm × 15 cm and applied homemade perilla oil to a JK wiper (manufactured by Nippon Paper Crecia Co., Ltd.). We then applied the oil completely to both sides of the Kraft paper to create oil-coated paper (Yokoyama et al., 2023). As natural drying of perilla oil takes several days, we dried the oil by heat, treating it at 120 °C for 135 minutes. We then soaked the prepared paper in water for three minutes and evaluated the water resistance of the paper based on the difference in weight before and after soaking, which we defined as the amount of water absorbed by the paper (hereafter referred to as "absorption").

As a result of applying perilla oil to the Kraft paper, the absorption was reduced to about 1/5 compared to uncoated Kraft paper, successfully imparting water resistance to the paper (see Fig. 7.3). However, it was also found that there were significant challenges. If the amount of perilla oil applied is below a certain level, water

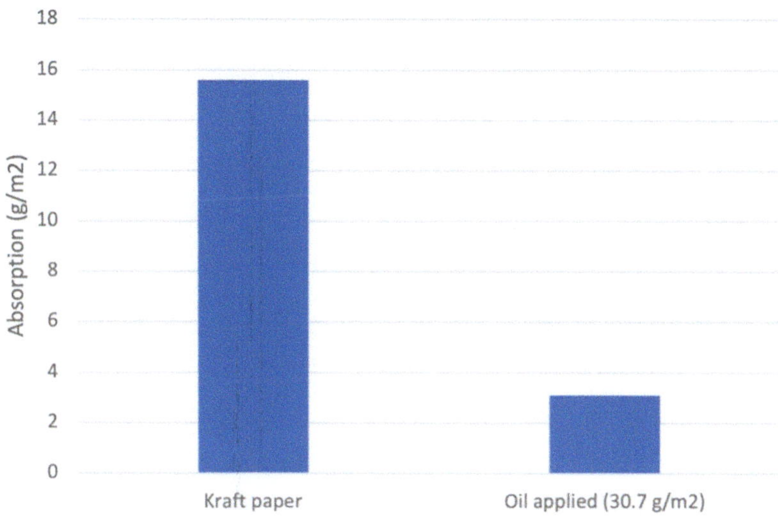

Fig. 7.3 Absorption of oil-coated paper

resistance cannot be achieved, and the absorption significantly increases (Fig. 7.4). In other words, there is a fixed amount of oil required per unit area and thickness to impart water resistance, and as the area and thickness of the paper increases, the amount of oil required increases proportionally, significantly increasing the cost. Furthermore, drying oils are generally expensive. For products like paper products that have a low unit price, this is a serious problem.

In the production of water-resistant paper by oil application, it is important to reduce the amount of oil applied as much as possible while maintaining low absorption. To achieve this, we focused on preventing the paper from absorbing the oil, keeping the oil only on the surface of the paper.

While walking and thinking of a good solution, Yokoyama, one of the authors, noticed water droplets forming perfect spheres on a leaf. Immediately, she thought, "This is it!" and considered mimicking the leaf's water-repellent properties. We then focused on the cuticle layer found in plant leaves shown in Fig. 7.5. The cuticle layer is a lipid film that covers the surface of plant leaves and stems (Hama et al., 2019). It has a water-resistant property and plays a role in protecting leaves and stems. This raises a question. Why do leaves, which, like paper, are primarily composed of cellulose, form a layer of lipids, or broadly speaking, oil on their surface? In fact, the cuticle layer has a multilayer structure, and there is a layer containing polysaccharides beneath the lipid layer (Hama et al., 2019). From this, we hypothesized that by creating another layer (hereafter referred to as "the first layer") between the paper and the oil, we could form an oil layer (hereafter referred to as "the second layer") on the surface of the paper and reduce the amount of oil applied.

A similar technique exists in the process of making traditional Japanese umbrellas which are made of *washi,* Japanese paper. In the case of Japanese umbrellas, Kakishibu is applied to *washi* before the application of drying oil (Imai, 2003).

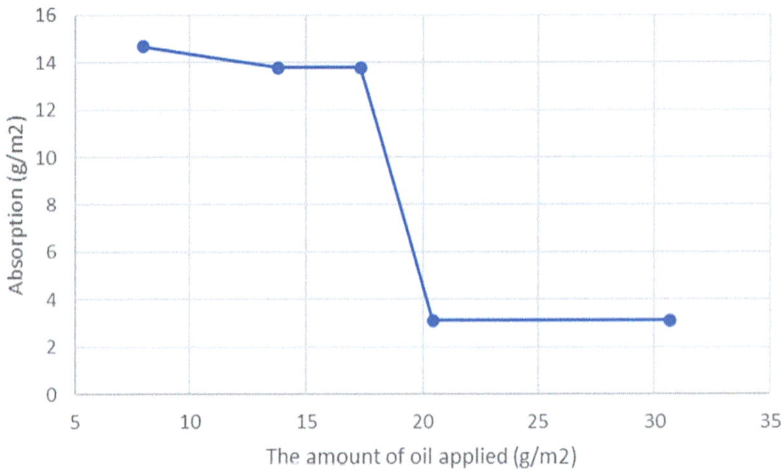

Fig. 7.4 Dependence of absorption on amount of oil applied

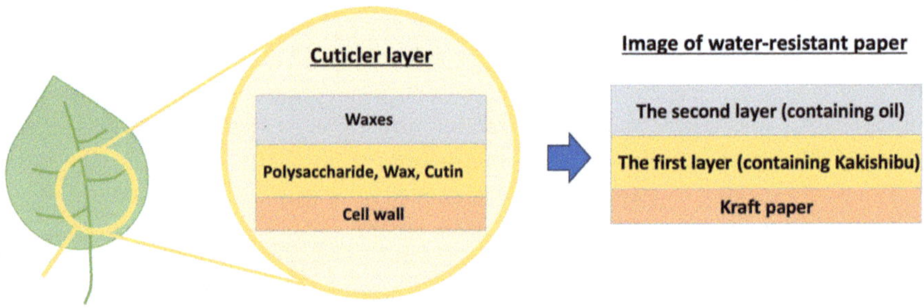

Fig. 7.5 Image and model substance of the developed water-resistant paper

We attempted to reduce the amount of oil applied by using Kakishibu (produced by Mimasu Kashichi Shouten Co., Ltd.) as the first layer material, following this traditional production method. This Kakishibu was applied to Kraft paper in the same way as the previously described oil application method and by drying it at 170 °C for 60 minutes (Yokoyama et al., 2023). Perilla oil was applied and dried on top of that paper, forming the second layer.

When the water absorption of this paper was measured, the result was 8 g/m^2 for paper with 16.4 g/m^2 of oil applied to the second layer. Considering that even when only oil was applied, the water absorption was 13.8 g/m^2 with 17.3 g/m^2 of oil applied, we were successful in reducing the amount of oil applied while maintaining the water resistance of the paper through applying Kakishibu as the first layer. Furthermore, when oil is applied to Kraft paper without any coating, the paper absorbs the oil, becomes transparent, and loses its texture. However, when Kakishibu is applied as the first layer, the paper does not become transparent even when oil is applied. When oil seeps into the paper, the air layers in the paper are filled with oil.

The difference of refractive index is larger between oil and cellulose than for oil and air. Therefore, when oil seeps in, the scattering of light within the paper is suppressed and light transmittance increases. From this, it is highly likely that by applying Kakishibu as the first layer, the paper was prevented from absorbing the oil, and as intended, a layer of oil was formed on the surface of the paper.

Let's return to the structure of the cuticle layer of a leaf for further inspiration and ideas. In the case of the cuticle layer, the layer beneath the lipid layer contains polysaccharides (Hama et al., 2019). Therefore, we thought that if we added sugar components to the first layer, the Kraft paper would become even less absorbent to oil, and we could impart water resistance with even less oil. We used a sugar binder (produced by Gun Ei Chemical Industry Co., Ltd.) and mixed this and Kakishibu in equal amounts to make the first layer coating (Yokoyama et al., 2023). The method of making water-resistant paper was carried out in the same way as the paper coated with Kakishibu, and the water absorption was measured. Turning now to the amount of perilla oil applied to the second layer, Fig. 7.6 shows the results of the amount of oil applied to the paper coated with Kakishibu as the first layer and the paper coated with a mixture of Kakishibu and sugar binder (Yokoyama et al., 2023). Both showed the same water absorption value of 8 g/m^2, but the amount of oil applied was about 8% less for the paper with the mixed coating as the first layer. By adding sugar components to the first layer, we were successful in reducing the amount of oil applied as expected, and we were able to create paper closer to the desired performance.

However, there are still many unknowns about the mechanisms shaping water repellency. From detailed observations with EDX (Energy Dispersive X-ray Fluorescence Spectrometer) and micro-IR (Micro InfraRed Spectrometer), we imagine that the layer structure we envisioned may have been formed, but we have not yet obtained reproducible data. At present, it is not clear why the application of

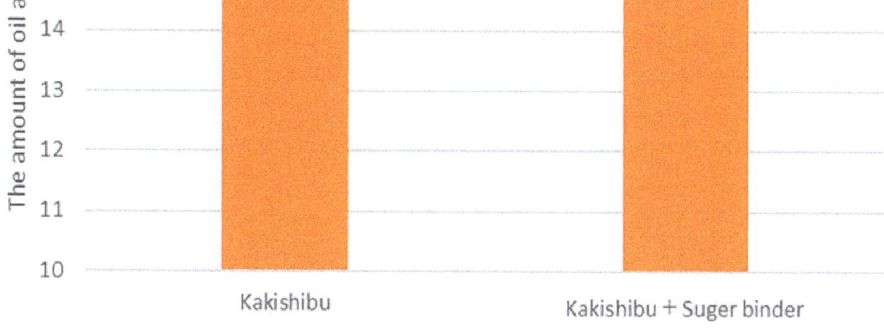

Fig. 7.6 The amount of oil applied on paper with first layer

Kakishibu or sugar binder can suppress the absorption of oil by the paper. There could be several reasons, but one major cause is likely that the Kakishibu we use is a natural material, and its components are not clearly known, making analysis difficult. This seems to be a common problem when trying to use traditional techniques. Regardless, if we can overcome these analytical challenges, we may be able to find more suitable materials for the first layer.

Nevertheless, we were able to develop water-resistant paper using natural materials, inspired by plants (cuticle layer) and traditional crafts (traditional Japanese umbrellas). Combining knowledge from different fields to advance the development of more environmentally friendly water-resistant paper is a significant achievement.

7.5.2 The Possibility of Mass Production

In the previous section, we examined a method of applying a primary coating of Kakishibu or a mixture of Kakishibu and sugar binder on Kraft paper, drying it, and then applying a secondary coating of perilla oil to realize a structure with a lipid wax layer on the surface. However, with the two processes separated, coating and drying, each drying condition being 170 °C for 60 minutes, it was difficult to move towards commercialization. Therefore, we examined conditions that could enable commercialization and mass production by reducing the number of coatings, such as through mixing coatings and shortening the drying time (Taniguchi et al., 2023, 2024).

First, we reviewed the water resistance test method. The water-resistant paper samples used were those using a mixture of Kakishibu and sugar binder as a base, as described in Sect. 7.5.1. The experimental method in Sect. 7.5.1, where the sample is submerged in water for three minutes, may allow water to seep into the paper from the end face of the paper. However, the cup test (using a Japanese industrial standard "JISP8140:1998"), a paper water resistance test, does not consider water seepage from the end face, so the results may differ from the standard test.

To evaluate the water resistance of the paper in a method closer to the industrial standard, we changed the experimental method to measure the amount of water absorbed by placing a sponge of 60 mm × 105 mm on the surface of a 150 mm × 150 mm paper sample for 60 minutes. At this time, the sponge was sufficiently soaked in water until no air came out when squeezed after being submerged in water. As shown in Table 7.1, it was found that the conventional method of submerging the paper in water was affected by the amount of water which is larger amount than from sponge (only from surface) that means water was absorbed from the end face of paper. Therefore, the new method was adopted for subsequent water absorption measurements.

The following sub-sections will cover two research and development processes: (1) reducing the number of coatings by emulsifying dispersion and (2) increasing the speed of coatings by UV irradiation, both leading to lower costs.

Table 7.1 Results of the review of the water resistance evaluation method

		Submerged in water		Using sponge	
Ingredient (%)	Kakishibu	0	100 (base coat)	0	100 (base coat)
	Mayonnaise		100 (base coat)		100 (base coat)
	2-propanol		100 (top coat)		100 (top coat)
	Perilla oil		100 (top coat)		100 (top coat)
Heat treatment conditions	Temperature(°C)	–	170	–	170
	Time (min)		60X2		60X2
Application amount; solids (g/m^2)*1		0.00	6.67/9.33	0	6.22/14.67
Immersion time in water (min)		3	3	60	60
Water absorption (g/m^2)		27.11	8.44	11.11	1.33
Water absorption reduction rate(%)*2		68.85		88.00	

*1 Base coat/top coat *2 Treatment effect of coating

1. *Reduction of the number of coatings by emulsifying dispersion*

   To reduce the number of coatings to one process, it is necessary to make a single liquid mixture of water-based Kakishibu, sugar and oil. A practical method to combine oil and water into a single liquid is to emulsify them. Because we are particular about using food as a material for the coating, we investigated commercially available emulsified foods and found that mayonnaise may be suitable. Mayonnaise is a food that emulsifies vegetable oil with the natural emulsifier "lecithin" contained in eggs and may be effective for achieving emulsification that is inexpensive and environmentally friendly. The materials used were:

- Kakishibu: Manufactured by Mimasu Kashichi Shouten Co., Ltd.
- Mayonnaise: Manufactured by Kewpie Corporation.
- 1 , 2-Propanol. Manufactured by FUJIFILM Wako Pure Chemical Corporation
- Perilla oil: Manufactured by Ota Oil Co., Ltd.
- Sugar binder Manufactured by Gun Ei Chemical Industry Co., Ltd.

We first attempted to emulsify the coating by making a coating using mayonnaise with the formulation in Table 7.2, putting it in a sealed container and shaking it vigorously. However, the water-based Kakishibu and the oil component of mayonnaise separated over time (several hours to one day), making it unstable as a coating. Therefore, we attempted to improve dispersion by applying an ultrasonic treatment (40 °C for 30 minutes) after mixing the paint. The ultrasonic treatment was performed using BRANSON (Yamato Scientific Co., Ltd.). As a result, as shown in Fig. 7.7, the dispersion improved with ultrasonic treatment, and it was possible to use it even after one week.

Next, we quantified the particle size before and after ultrasonic treatment by particle size distribution measurement. The particle size distribution measurement was performed using the Laser Diffraction Particle Size Analyzer LA-920 made by HORIBA, Ltd., and the liquid was circulated for measurement. Results are shown in Figs. 7.8 and 7.9. In the untreated paint, the average particle size was 32 μm, whereas in the ultrasonically treated paint, the average particle size was 10 μm,

Table 7.2 Results of the water repellency with and without mayonnaise conditions

Mayonnaise		With	Without
Ingredient (phr)	Kakishibu	100	100 (base coat)
	Rasitop	100	100 (base coat)
	1-propanol	0	100 (top coat)
	Perilla oil	100	100 (top coat)
	Mayonnaise	100	0
Heat treatment conditions	Temperature (°C)	170	170
	Time (min)	**60 × 1**	**60 × 2**
Application amount; solids (g/m^2)*1		15.11	6.67/9.33
Immersion time in water (min)		60	60
Water absorption (g/m^2)		2.22	1.33
Water repellency (JIS P8137)		R10 (the highest water repellency)	R10 (the highest water repellency)

Fig. 7.7 Appearance of the paint after one week

indicating a reduction in particle size. The cumulative frequency (%) of particles larger than 67 μm also exceeded 10% in the untreated paint. However, after ultrasonic treatment, the cumulative frequency was 0.2% or less in the ultrasonically treated paint, indicating a decrease in the proportion of larger particles.

From these results, it was concluded that the stability of the paint can be improved by reducing the particle size of the emulsion with ultrasound.

Next, we examined the formulation using mayonnaise. First, when using a sugar binder, it is necessary to wait for the curing reaction to proceed. Therefore, a drying time of 60 minutes at 170 °C was essential. Furthermore, there are problems such as the paper becoming brittle when using a sugar binder. Therefore, when considering the formulation using mayonnaise, we conducted a study without using a sugar binder. The results are shown in Table 7.3. Without mayonnaise, two coats—base

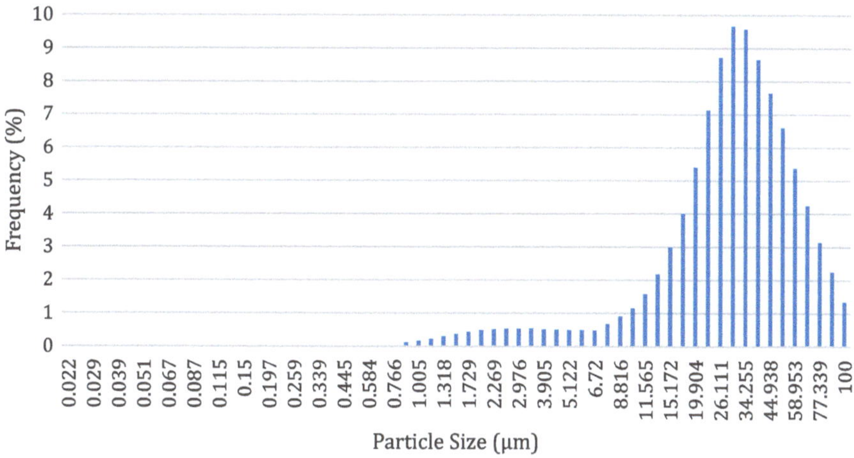

Fig. 7.8 Particle size distribution without sonic wave treatment

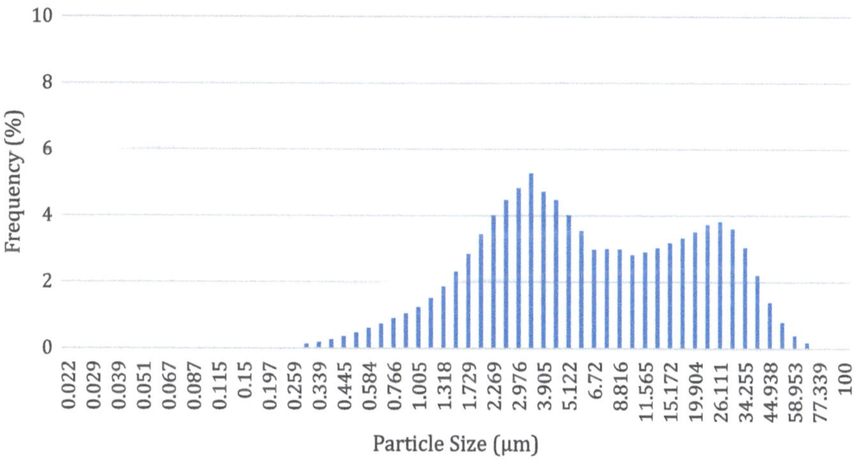

Fig. 7.9 Particle size distribution after sonic wave treatment

and top—are required. However, when mayonnaise is used, all materials can be applied in a single coat, providing both water resistance and water repellency. We were successful in combining two types of paints into one and reducing the drying time from 60 minutes $\times$ 2 times (without mayonnaise condition in Table 7.3) to 60 minutes × 1 time (with mayonnaise condition in Table 7.3).

It was hypothesized that the number of paint applications could be reduced to one with the following explanation. First, when the dispersion of the emulsified Kakishibu-mayonnaise paint is poor, it separates into Kakishibu (lower layer) and oil (upper layer). This paint separates oil as the balance of the emulsion is disrupted by the evaporation of water and 1-propanol during the process of drying on Kraft paper. As a result, we imagine that it can simulate the cuticle layer devised in 6.5.1

Table 7.3 Results of shortening drying conditions study

Sample no.		1	2	3	4
Ingredient (%)	Kakishibu	100	100	100	100
	Mayonnaise	75	75	75	75
	Perilla oil	100	–	–	–
	2-propanol	75	75	75	75
Heat treatment conditions	Temperature (°C)	170	170	170	170
	Time (min)	60	60	15	5
Application amount; solids (g/m^2)		11.11	11.11	6.22	5.78
Aging at room temperature (days)		0	0	7	7
Water absorption (g/m^2)		2.66	2.66	7.56	8.00
Number of applications		1	1	1	1
Water repellency (JIS P8137)		R0	R8	R8	R8

by separating into an oil layer on the surface side of the Kraft paper and a Kakishibu layer on the substrate side.

Additionally, from these results, it was found that the water repellency increases if only mayonnaise is used without using perilla oil. This is thought to be because the vegetable oil derived from mayonnaise, which has been processed into smaller particles, showed higher water repellency than the larger particles of perilla oil. However, this mechanism remains unknown.

On the other hand, in the case of a drying time of 5 minutes, water repellency was imparted only when it was stored in the air for about one week. Being able to reduce the drying time to 5 minutes is a welcome thing when considering the mass production process. However, in a conventional paper production line, the paper is wound up in a roll and shipped. In this case, the paper inside the roll does not come into contact with air, so water repellency does not appear even if it is stored for a long time. Therefore, it is necessary to have water repellency immediately after drying.

2. *Increasing the speed of coatings by UV irradiation*

In traditional Japanese umbrella production, it is dried in the sun outdoors after applying Kakishibu. Since the chemical structure of Kakishibu contains many conjugated structures of π electrons, we thought that UV might be a factor. Furthermore, the components of Kakishibu undergo oxidation reactions in the presence of oxygen (Shimamoto, 2016). We hypothesized that the water repellency imparted over time is due to the polymerization reaction of Kakishibu. The effect of traditional sun drying may also be due to the promotion of oxidation reactions by UV.

In order to check the hypothesis that the polymerization reaction of Kakishibu is promoted by UV, we attempted to expose Kraft paper, which had been coated with paint before drying, to UV light. The UV irradiation device used was a 1.5 kW UV lamp manufactured by Eye Graphics Company. The light intensity was measured using a UVPF-36 illuminometer from Eye Graphics Company, and the line speed was adjusted to achieve the desired irradiation amount.

The results are shown in Table 7.4. After irradiating with 500–1000 mJ/cm^2 of UV using the UV irradiation device and then drying, the water repellency was

Table 7.4 Results of UV irradiation conditions study

Accumulated light quantity (mJ/cm^2)	500	750	1000	1000	1000
Drying temperature (°C)	170	170	170	120	120
Drying time (min)	5	1	0.5	1	0.5
Water repellency (JIS P8137)	>R9	>R9	>R9	>R9	>R9

confirmed. As a result, it was confirmed that water repellency can be imparted immediately after drying, even in drying at 120 °C for 0.5 minutes. The UV irradiation successfully shortened the drying conditions, which was perceived as a challenge for mass production. The water-resistant paper used here had a solid content coating amount containing oil that was 1/5 of the coating amount when only perilla oil examined in 6.5.1 was coated. This result allows for a significant reduction in the amount of oil to be coated. Furthermore, the oil contained in mayonnaise is either non-drying oil or semi-drying oil, so there is no need to use expensive drying oil. As such, it is possible to produce water-resistant paper at a price that is orders of magnitude cheaper than the original formulation and manufacturing conditions we were considering. On the other hand, we are investigating the mechanism of water repellency expression using Fourier Transform InfraRed Spectroscopy (FTIR) and Electron Spin Resonance (ESR), and we observed that oil increased on the paper surface, but we have not yet obtained reproducible data. Further improvements in characteristics and production efficiency will require further investigation.

In any case, it has become possible to express water repellency in about 0.5 to 1 minute by irradiating with UV. Although there are still issues such as long drying times for mass production, these results are at a level that allows for prototyping on a mass production line, indicating the potential for future industrialization and commercialization.

7.6 Background Story of Internship Management: Inspired by the Minakata Mandala Rooted in Buddhism

Finally, we would like to introduce how to manage the interns by Goto. The management method was unique in that he introduced a Buddhist worldview. Therefore, the management was different from a typical Western one including the process of reporting to superiors, preparation for formal meetings, creation of standard documents, management of research budgets, negotiations within and outside the company.

Goto has already applied the same management to the development of vacuum-formable polycarbonate foam sheets in parallel with this internship. In this project, we were able to realize in two years, from setting up a prototype line in the lab to making a prototype that can submit samples to customers by forming a practical level sheet of 1 m width and reaching a state where small lot production is possible with a team of about six people. Normally, it takes 3–5 years to set up a sheet forming line, and if you are going to make a foaming sheet that does not exist in the world like this, it takes more time. He thinks that the fact that we were able to advance such development in two years is a result that supports this way of management.

The foundation of the management is the philosophy of Minakata Kumagusu (1867–1941), who was a biologist and a Buddhist scholar devoted to the study of slime molds. Minakata is also known as the Japanese author who has published the highest numbers of papers in Nature journal. In an exchange of letters with a high priest of Koyasan, Horyu Dogi, he writes about his Buddhist view of the universe, asserting that it does not contradict modern science (Minakata, 2023). His letter does not mention the use of this cosmology in real-world management, but it does present a way of thinking that comprehensively deals with everything from matter, people, and animals to metaphysical concepts. Goto finds this way of thinking helpful in promoting research and development.

Here, we would like to discuss mandala, which is a symbolic representation of the universe in Ancient Hindu and Buddhist philosophy (Matsui, 2007), proposed by Minakata (called Minakata mandala), and think about research and development management after considering the relationship between research activities and the mandala. Mandala includes lot of original concepts from Minakata and Buddhism, and English interpretations are based on the research of Matsui (2007) and Nakanishi (2019).

The concept of Minakata mandala is shown in Fig. 7.10. There is the Diamond world (*Kongo Dainichi*) on the right and the womb world (*Taizo Dainichi*) on the left (see Fig. 7.10). The Diamond world comprises the laws and theorems that cover the entire world, including the outside of the universe. This includes laws and theorems that are not recognized by human beings. In this, there is wisdom and cognition. From this Diamond world, the "mind" separates and becomes "substance." For example, a stone is a thing and also something that a person recognizes as a stone, but a stone lying on the street is just a stone as a "substance." The "mind" and

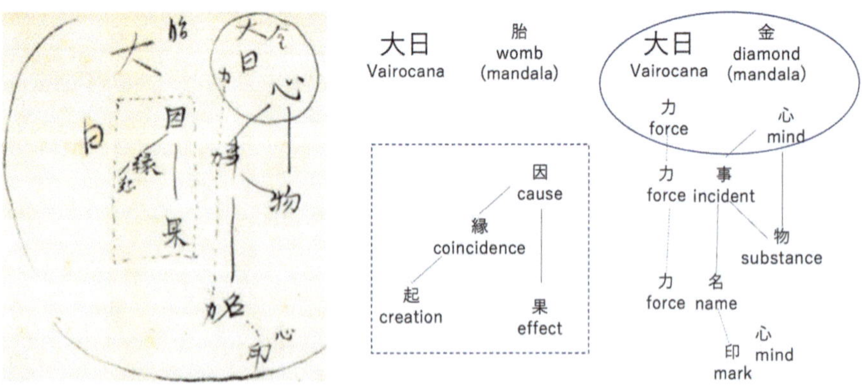

Fig. 7.10 Minakata's Image of the Universe (Mandala) (left) and its translation by the authors (right). (*Source* (modified with permission): Minakata (2023))

“substance” interact to create “incident.” Someone thinking that a stone is beautiful is such an “incident,” and some kind of will power or force is at work in causing such an “incident,” although not everyone will think the same.

Things that just meet or are nearby are “Coincidence,” and the act of things interacting to cause something is called “*engi*,” creation. “Creation” becomes a “name” by happening repeatedly. For example, in Kyoto, there are seven named stones called Kamo Shichiseki (the Kamo 7), each with its name because someone must have thought these stones were beautiful. Using them as garden stones, the “coincidence” of various people thinking a stone is beautiful contributes to the becoming of a “name.” As the “name” spreads in society, the “name” or “collection of names” leaves some kind of impression on people. This action on the mind is called a “mark.” It is the impression that the Kamo 7 stones are high-end garden stones.

On the other hand, everything in this world, “incident,” “substance,” and “mind” are made up of causality, “cause and effect.” It refers to all phenomena, such as the universe being born in the Big Bang and disappearing, the sun being born and disappearing, people being born and dying, anger arising and subsiding, ice forming in winter and melting and disappearing in spring. Figure 7.11 shows these beginnings and ends and the process of beginning and ending schematically with a line to represent various causal relationships. Moreover, these lines also represent spatiotemporal trajectory of “all things in the universe (ATU),” macro and microscopic things and their natural law, conept, like elementary particle, atoms, molecules, materials, anti-materials, life, animal, society, trib, nation, star, starsystem, galaxy and universe et. ct.

Fig. 7.11 is a highly abstract diagram that may not be immediately understandable. Therefore, we will add annotations ① through ⑥ and provide an explanation using concrete examples.

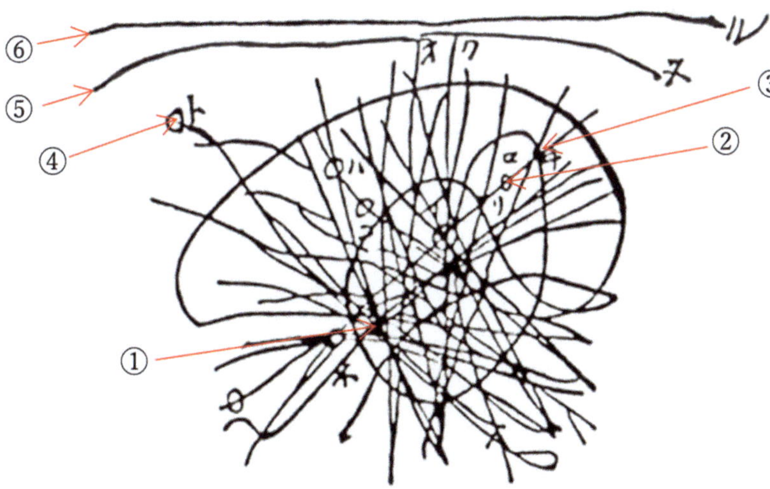

Fig. 7.11 Image of events and reasons filling the universe (cause and effect). Red arrows and numbers were added by the authors. (*Source* (modified with permission): Minakata (2023))

When looking at Fig. 7.11, think of a person at the center, along with "incident," "substance," "mind," and the ATU existing in time and space. It is easy to find the ATU for the point of ① where many of the ATU intersect. Although the point ② does not intersect with any other ATU at that location, there are points such as ③ nearby which it does intersect with and ② can be seen reaching other ATUs that connect to point ③. Point ④ has little involvement, but its ATU intersects with other ATU, and the point of ④ can be reached in this way. The ATU of ⑤ touches other ATU very slightly, and it is difficult to find them, but they can be found from the events and reasons affected by other ATU at the contact point. The point of ⑥ does not intersect with any ATU at all, but it is close to ⑤. So, ⑥ indirectly affects ⑤ and can barely find such affection in ⑤ that shows the existence of its events and reasons of ⑥ through ⑤.

Examples of point ① include people meeting people, something hitting something else and breaking, and molecules hitting molecules that cause chemical reactions. Point ② could refer to people who do not meeting at this point but just before or after meet with or aware with other people at point ③. Point ④ could relate to death or birth because it is beginning or end of trajectory. So, this point represents cause or effect of Fig. 7.10. So, Fig. 7.11 is not the perfect picture of universe because cause is created by creation that means that there are other ATU around the point ④. The line ⑤ examples the facts that "the earth is moving, the pathogen is a virus," or "Mr. A is ruining the atmosphere, and the teamwork is disrupted." Line ⑥ could represent theories and phenomena that people cannot directly experience, such as the Big Bang, and quantum theory can represent theories and phenomena that people cannot directly experience, such as the Big Bang and quantum theory.

Each encounter between ATU is termed a "coincidence" that the encounter of such principles and phenomena. A coincidence does not necessarily lead to a realization or a major source of change. Only when there is affection like realization or contact does the principle interact with each other, causing change and becoming "creation." For example, coincidence represents the situation that two strangers sit next to each other but failing to greet. Creation represents the situation that two strangers sit next to each other noticed and say hello to the other that cause the conversation or friendship. The following provides more examples and explanation:

Example 1: Someone noticed ("force") the stone ("substance") and moved it ("force"). In this case, "force" resulted in the movement of the stone. The relationship between the stone and someone can be interpreted as "coincidence" and the fact that the stone moved can be interpreted as "creation."

Example 2: If someone is not interested in the stone and touches it, nothing will happen, even if someone walks next to the stone. So, "creation" only can happen when "force" is applied.

Example 3: While walking along the Kamo River ("incident"), you found ("force") a stone ("substance"). You thought it was beautiful ("force") and brought it home ("force"), placing it in your garden ("incident"). In this way, the "creation" of the garden with stones is made by the "force" of noticing them, the feeling of beauty, and the action of bringing them back to the home. It can be an "incident" of the creation of a garden with the brought back stone. The "force" is considered necessary to "cause," "name," and "mark" these incidents.

Let's reinterpret these cosmological views in the context of research and development activities. Various people are involved in such activities. From raw material suppliers to prototype contractors and future users, people are involved in all kinds of business flows. For researchers and developers in particular, relationships with objects are important. This involves seeing or touching the objects researchers deal with, which may spark interest or thought. For example, someone ("mind") noticed ("force") and talked to ("force") a colleague ("mind"), the colleague ("mind") found it interesting ("force"), and the other researchers who noticed ("mind") it proceeded with the verification ("force"); as a result, the reproducibility conditions were clarified and a journal article named "XX" ("name") was published. In other words, if "force" does not work when the "mind," "substances," and "incidents" are interacting complexly, no interaction occurs, which means "creation" does not occur. One of the goals of research and development activities in companies is creating products ("name"), and at a higher level, it is to create a market ("mark") for such products. Research and development activities are about turning coincidence into creation, creation into names, and names into marks. So, the most important thing in research and development management is how to manage "force" in the community. How can management make it easy for "force" to work?

Reactions between molecules require energy. In Minakata's view of the universe, "substances," "incidents," and "minds" exist in the same way, so thinking about the reaction between molecules also applies to the interaction between humans ("minds"), between humans ("minds") and nature ("substances"), and between "minds" and "incidents." What can be controlled to some extent are "minds." Therefore, how to effectively maintain the high energy of "minds" is key to management.

With respect to interns involved in research, it is important to provide limited guidance so that they can fully face the object on their own terms. This provides the opportunity to focus on development and creates a situation where the "force" in Minakata's sense of the universe is able to emerge. This may also be extended to outside internship research and development, such that "coincidence" may more readily become "creation."

Finally, we would like to provide three specific examples of internship management using Minakata Mandala.

First, when Yokoyama joined Maxeell as an intern, we initially discussed how to achieve water resistance paper using natural materials. During that conversation, Goto suggested that the only way might be to observe what exists in nature and try to mimic it. Goto also told Yokoyama to take a walk on a rainy day and think about it. Later, Yokoyama came back and reported that she had noticed camellia leaves on a rainy day and thought they might be a useful reference. This observation became the foundation for the concept shown in Fig. 7.5. In this case, Yokoyama walking in the rain corresponds to state ② in Fig. 7.11. The prior conversation Goto had with her represents state ③. By noticing a different ATU namely, the camellia leaf—during state ②, the concept in Fig. 7.5 was formed. Camellias are commonly planted throughout Japanese cities, and Yokoyama must have seen them before. However, she had not arrived at that idea during our initial discussion. It was only after our conversation (incident) that Yokoyama was able to recognize ("force") the water-repellent nature of the leaf (incident). By investigating the structure ("substance") of the leaf ("force"), Yokoyama was able to propose ("force") the concept ("incident").

Second, we had discussed the need to combine the separate steps of applying persimmon tannin and oil. Goto gave Taniguchi the following advice: since persimmon tannin is water-based, the only way to mix it with oil in a single solution would be to emulsify it. That led to the idea of looking at emulsified food products available in supermarkets. Goto gave him some pocket money and sent him off. After several weeks, he came back with mayonnaise. The experiment using this became the basis for the formulation concept of the coating currently under development. In this case, Taniguchi walking alone through the supermarket corresponds to state ②. The prior discussion about emulsification represents state ③. While exploring the supermarket in state ②, he used the small amount of money Goto had given him ("force") to buy ("force") mayonnaise ("substance") and used it in an experiment ("force"). As a result, Taniguchi was able to create a water-repellent coating ("incident"), which led to a patent application ("name").

Finally, we also share the background behind the idea of using UV. This episode occurred when Yokoyama was already working at Maxell and Taniguchi was there as an intern. We were discussing how to solve the problem of delayed water repellency after applying the coating. At this time, discussion with 3 person corresponded to state ① that persons and information are gathered in one place. At this meeting, "traditional Japanese umbrellas" has arisen, and the idea of UV that is not directly connected to our mind connected to "UV." In this case, UV refers to the state ⑤ that is invisible and hardly to feel except sun burn of skin. In this case, Taniguchi ("mind") remarked that the manufacturing method resembled that of traditional Japanese umbrellas ("force"). Hearing that, Goto pointed out that traditional Japanese umbrellas are typically dried in the sun ("incident"). On hearing this, Yokoyama ("mind") suggested trying UV irradiation ("force"). Goto then investigated the equipment available in-house ("force") and found a UV irradiation device ("substance"). Taniguchi ("force") conducted an experiment using it, and as previously shown, we discovered that UV irradiation could rapidly induce water repellency ("incident"). This technology also led to a patent application ("name").

All of these cases, you can see that many forces were at play an important role. In this way, by adopting Minakata's way of thinking, being mindful of "forces," and creating conditions in which those forces can easily act, it is possible to accelerate research and development.

7.7 Conclusion

We discussed traditional technologies in relation to corporate activities and the prospects for development in this chapter. We explored the difficult situation regarding the use of company patent data and issues to do with budgets. To overcome this latter issue within the company, we adapted and relied on interns for advancing technical development. Sensibilities from Eastern thought is adopted, which represent the universe including material to human thought that is called Minakata Mandala, originally derived from Buddhism, to inform research management approaches. We also explored attempts to develop water repellent paint for paper, along with efforts to improve efficiencies and productivity with respect to the process and materials. Going forward, we will likely have to start with a volume that does not allow for cost reduction through scaling up, and so we expect continued challenging circumstances. Without being disheartened by this, we aim to continue using our scientific knowledge to strive for further enhancements and efficiencies, flexibly incorporating traditional and Eastern insights, and involving not only academia and industry, but also farmers and local governments, in our efforts towards societal implementation.

Other issues include current limits to the production capacity of Kakishibu, which will need to be addressed through the revival of regional agriculture. In addition, we should be concurrently exploring tannins and polyphenols that could replace and serve as alternatives to Kakishibu. Regardless, moving forward from here requires the development of scalable scenarios, securing budgets based on these scenarios, concretizing their applications, and cultivating customers.

Ultimately, in order to make this technology a successful example of the application of indigenous knowledge to the modern world, more and more people who consider the potential of traditional knowledge to environmental issues are needed toward green innovation with Kakishibu. So, we hope this example of industrialization of the indegenous technology and understanding of management using eastern knowledge, stimulate the peaple who engaged in not only industry and academia but also consumer, politician and all peple who are living in the modern world.

Acknowledgments We would like to express my gratitude to the general affairs and intellectual property personnel at Maxell, Ltd. for being able to conduct this research as part of the experience of accepting interns and undertaking entrepreneurial research.

We also would like to express my gratitude to the secretariat Mie Tanaka, Professor Atsushi Tani and Professor Nobuko Oomido of Kobe university, where the interns belong for their understanding, which allowed us to implement this irregular internship program.

References

Administrative Evaluation Bureau, Ministry of Internal Affairs and Communications. (2022). Dento kogei no chiiki shigen tositeno katsuyo ni kansuru jittai chosa kekka hokokusho (Report on the survey of the utilization of traditional crafts as regional resources).

Ali, W., Ali, H., Gillani, S., Zinck, P., & Souissi, S. (2023). Polylactic acid synthesis, biodegradability, conversion to microplastics and toxicity: A review. *Environmental Chemistry Letters, 21*, 1761–1786. https://doi.org/10.1007/s10311-023-01564-8

Bloom, N., Jones, C. I., Van Reenen, J., & Webb, M. (2020). Are ideas getting harder to find? *American Economic Review, 110*(4), 1104–1144. https://doi.org/10.1257/aer.20180338

Edamura, K., Miyagawa, T., & Uchiyama, K. (2017). Kankyo bunya ni okeru keieishigen no chikuseki to kigyokachi: Kankyotoushi katsudo kara mita jissho bunseki (Accumulation of management resources in the environmental field and corporate value: An empirical analysis from the perspective of environmental investment activities). *RIETI Discussion Paper Series*, 17-J-027.

Gordon, R. J. (2016). *The rise and fall of American growth: The US standard of living since the civil war*. Princeton University Press.

Goto, K., & Takashima, C. (2021). Dento kogei niokeru bunka to sangyo no jizokusei (Sustainability of culture and industry in traditional crafts). *21 seiki ni okeru jizoku kanona keizaishakai no sozo ni mukete (Toward the Creation of a Sustainable Economy and Society in the 21st Century), 1*(4), 11–20.

Hama, T., Seki, K., Ishibashi, A., Miyazaki, A., Kouchi, A., Watanabe, N., Shimoaka, T., & Hasegawa, T. (2019). Probing the molecular structure and orientation of the leaf surface of *Brassica oleracea* L. by polarization modulation-infrared reflection-absorption spectroscopy. *Plant & Cell Physiology, 60*(7), 1567–1580. https://doi.org/10.1093/pcp/pcz063

Hosokawa, J., & Nishiyama, M. (1990). Tennenn kobunshi no zairyoka (Materialization of natural polymers). *Polymer, 39*(4), 276–279.

Imai, K. (2003). *Kakishibu: Mono to Ningen no Bunkashi (Kakishibu: Cultural history of things and humans)*. Hosei University Press.

Matsui, R. (2007). Kindai kagaku no Togo riron toshite no mandara (Mandala as a synthetic theory of modern sciences). *Kokusai bunka kenkyu (International Cultural Studies), 11*, 29–41.

Minakata, K. (2023). In S. Nakazawa (Ed.), *Minakata mandala* (8th ed.). Kawade Shobo Shinsha.

Miyazawa, T. (2018). Chemistry of natural and synthetic fibers. *Chemistry & Education, 66*(9), 444–447.

Nakanishi, S. (2019). Minakata mandara no eiyaku (English translation of Minakata Mandala). *Kumagusu kenkyuu (Kumagusu Research), 13*, 290–302.

Nakanishi, T. (2022). Nihon no seizogyo niokeru kenkyukaihatsu no sentaku toshuchu to sono koka ("Selection and concentration" on research and development and its effect in the Japanese manufacturing industry). *Kan taiheiyo daigaku kenkyu kiyo (Pacific Rim University Bulletin), 20*, 49.

Shekhar, N., & Mondal, A. (2024). Synthesis, properties, environmental degradation, processing, and applications of Polylactic Acid (PLA): an overview. *Polymer Bulletin*, 81, 11421 - 11457. https://doi.org/10.1007/s00289-024-05252-7

Oda, K. (2014). Materials research on Kakishibu-Kamiko and its uses in dress design. *Journal of the Japan Research Association for Textile End-Users, 55*(5), 41–46.

Orihata, K. (2006). Selection and focus of the businesses in the restructuring process of the corporations. *Tama University Journal of Management and Information Sciences, 10*.

Patent Office of Japan. (2003). Shokumu hatsumei ni kakawaru gaikoku tokkyoken to no atsukai ni tsuite (Handling of foreign patent rights related to job inventions).

Paul, C. S., Stefanovic, L., Robledo-Mahón, T., Mercl, F., Száková, J., & Tlustoš, P. (2024). Agricultural waste-based lactic acid production by the fungus *Rhizopus oryzae*: A tool for sustainable polylactic acid production for agricultural use—A review. *Plant, Soil and Environment, 70*(12), 739–750. https://doi.org/10.17221/416/2024-PSE

Shimamoto, T. (2016). The chemistry of kakishibu, deeply rooted in Japanese culture. *Chemistry & Education, 64*(7), 348–349.

Shimono, A. (2003). *Chuusho kigyou kenkyuu (Small Business Research), 75*, 11–77.

Taniguchi, S., Morioka, M., Yokoyama, Y., & Goto, T. (2023). *Composition, coating, waterproofed substrate, and method for producing the waterproofed substrate*, (PCT patent application number, PCT/JP2023/044632), Applicants: Maxell, Ltd.

Taniguchi, S., Morioka, M., Yokoyama, Y., & Goto, T. (2024). *Method for producing waterproofed cellulose fabric paper* (Japanese Patent Application Publication No. 2023-210266). Applicants: Maxell, Ltd.

Wang, Y., & Takahashi, K. (2020). Kenkyu kaihatsu toushi to innovation (Research and development investment and innovation: An approach using patent data). *Nihon ginkou waakingu peipaa shiriizu (Bank of Japan Working Paper Series)*, no. 20-J-2.

Yamauchi, T., & Usami, N. (2005). Characteristic mechanical properties of washi papers. *Japan Technical Association of the Pulp and Paper Industry Journal, 59*(10), 104–116.

Yokogawa, M., & Nakano, K. (2016). *Social and economic value of Japanese traditional crafts*. Master thesis at the Graduate School of System Design and Management, Keio University.
Yokoyama, Y., Goto, T., & Ishii, S. (2023). *Cellulose cloth paper impregnation agent kit, waterproofed cellulose cloth paper, and its production method* (Japanese Patent Application Publication No. 2022-28306). Applicants: Maxell, Ltd.; Gun Ei Chemical Industry, Ltd.

Toshiharu Goto earned his bachelor's and master's degrees at Hokkaido University through research on the mechanical properties of bone. He then worked at Hitachi Cable, Ltd. and Nitto Boseki Co., Ltd., where he was engaged in the development of environmentally conscious technologies, such as polymer recycling and formaldehyde-free heat insulation materials. His research earned him awards from organizations including the Society of Chemical Engineers, Japan, and the Research Association For Feedstock Recycling of Plastics, Japan. He also obtained a doctoral degree in chemical engineering at Shizuoka University. Currently, he works at Maxell, Ltd., engaging in engineering and marketing related to polymer molding and water-resistant paper. In parallel, he is exploring ways to enhance creativity in manufacturing by applying the conceptual framework of the "Minakata Mandala."

Yuka Yokoyama earned her bachelor's and master's degrees in the study of Hydrogen transfer reactions in Chibaite at Kobe University. During her academic tenure, she engaged in an internship at Maxell, Ltd., focusing on the development of water-resistant paper. Currently, she works at Maxell, Ltd., engaging in development of new technology related to green innovation.

Shota Taniguchi earned his bachelor's and master's degrees from Kobe University through phylogenetic studies of plant "*hydrangea*" using on plant genome analysis and gene imaging techniques. During his academic tenure, he participated in an internship at Maxell, Ltd., contributing to the development of water-resistant paper. Following graduation, he joined Maxell, Ltd. as a full-time researcher, engaging in the design and development of all-solid-state batteries. Leveraging his expertise in materials science, he has been involved in the development of electrode materials and battery design aimed at the mass production of heat-resistant all-solid-state batteries.

Part IV
Culturally Relevant STEM Education on Kakishibu

Chapter 8
Effects of Culturally Relevant STEM Education: Children's Eco-Friendly Invention Ideas Using Kakishibu in Japan and Australia

Kei Kano and Graham Walker

Abstract In this chapter, we outline the influence of an experimental, hands-on Kakishibu STEM (Science, Technology, Engineering and Mathematics) workshop designed for primary school children and their families, guided by the principles of Culturally Relevant Pedagogy (CRP). The chapter commences with a literature review on CRP, with particular emphasis on STEM education. Subsequently, the development process of culturally relevant STEM education workshops centred on Kakishibu is introduced, incorporating CRP and a utility value intervention involving worksheets on self-generated invention ideas involving Kakishibu. The workshops were conducted in both Japan and Australia, where the relevance of Kakishibu and policies on single-use plastic bans vary. A qualitative analysis was performed on the collected worksheets from eco-friendly and sustainability perspectives. The findings reveal that the majority of children proposed eco-friendly applications for Kakishibu; however, the context of single use plastics in everyday life had a discernable effect. Additionally, some children advocated a systems approach to broader plastic issues. Consequently, it is suggested that the intersection of multicultural relevance and single-use plastic alternatives may produce a synergistic effect. Given that children employed an anthropocentric perspective rather than focusing on the foundational philosophy of Kakishibu, it is necessary to emphasise shifting the concept of "culturally relevant" from everyday life to the cultural philosophical background.

K. Kano (✉)
Faculty of Education, Shiga University, Otsu, Shiga, Japan

The Australian National Centre for the Public Awareness of Science, Australian National University, Canberra, ACT, Australia
e-mail: kkano@edu.shiga-u.ac.jp; Kei.Kano@anu.edu.au

G. Walker
The Australian National Centre for the Public Awareness of Science, Australian National University, Canberra, ACT, Australia
e-mail: graham.walker@anu.edu.au

K. Kano et al. (eds.), *Traditional Knowledge for Green Innovation with Japanese Persimmon Tannin, Kakishibu*, https://doi.org/10.1007/978-981-95-7805-4_8

Keywords Culturally relevant pedagogy · STEM · Hands-on · Informal education · Kakishibu · Traditional knowledge · Utility value intervention · Plastic · Sustainability · Drawings

8.1 Introduction

"Most people, even Japanese, do not know Kakishibu. Therefore, what we have to do first to promote Kakishibu is education." This is an essential voice we often hear in traditional knowledge, such as in Kakishibu. One of the backgrounds to this kind of voice is that science education in Japan has been oriented to teaching Western Science since the Meiji era. This is because Japan should have been catching up with Western skills (the details are in Chap. 2). For example, phenolic resin, the first petroleum-derived synthetic polymer in the world, is explained in detail in science textbooks in Japan, from its chemical structure to its usage in daily life. On the other hand, Kakishibu has not been covered in science textbooks. Kakishibu is a natural polymer of kaki-tannin and has over 1000 years of history. Bringing intangible cultural heritage together with an apt response to current environmental problems, one would think Kakishibu should have been compared with phenolic resins in science textbooks—contrasting the natural polymer versus synthetic plastic. Yet it remains conspicuously absent.

Moreover, even the background of Western Science, or Western philosophy, has not been taught in science textbooks in Japan. Science learning has more focused on collecting facts and theories, with less emphasis on the creativity, critical thinking, and scientific method. Rather, science education in Japan has largely functioned as a display case for Western scientific knowledge, even though Watanabe (1991) argues that because Western science was broadly accepted mainly through training and education, this situation should be corrected first in schools, beginning with a shift in attitudes toward Western learning. In addition, the decline in students' positive attitude toward science phenomenon has been pointed out since the late 1980s (Naganuma, 2015; Kano, 2023). The influence of *Wakon-Yosai* (a combination of Japanese tradition and Western skills), see the details in Chap. 2) on science education has been significant, as Watanabe (1991) expressed his regret as follows.

> How regrettable that education in science, instead of contributing to an understanding of modern Western thought or generating an enthusiasm for human achievement in interpreting nature has instead become a form of intellectual torture for pupils and students and a nightmare for grownups when they recall their school days.

We think it is time to expand the concept of science education in Japan to include both Western philosophy and Japanese views of nature (see the details in Chap. 2). In harmony, these will give students' and even teachers' a more agile, integrated understanding of science, technology, and society, along with the critical consciousness needed to address real-world problems and global socioscientifc challenges.

To achieve the expansion, we propose the development of culturally relevant STEM education based on culturally relevant pedagogy (CRP), which is "a pedagogy that empowers students intellectually, socially, emotionally, and politically by using cultural referents to impart knowledge, skills, and attitudes" (Ladson-Billings, 1994, p.17-18). This broader shift in perspective could be a thoughtful answer to the need for education to promote Kakishibu mentioned at the beginning of this chapter, because it would educate people with cultural competencies and critical consciousness going beyond just promotion. Even Japanese children no longer know about Kakishibu, despite it being traditional Japanese knowledge. Kakishibu is but one opportunity to use CRP to give students an integrated view of science, culture, history, and philosophy and gain critical skills to address real-world problems, such as addressing plastic waste through innovative—yet traditional—use of Kakishibu.

8.1.1 Literature on Culturally Relevant Pedagogy (CRP) in STEM Field

CRP leverages cultural elements to develop knowledge, skills, and attitudes across broad domains from the social and emotional to intellectual and political (Ladson-Billings, 1994). CRP rests on three major domains (Ladson-Billings, 1995, 2014). The first domain is academic success or student learning, which refers to intellectual growth that students experience as a result of classroom instruction and learning experiences. Second, cultural competence refers to the ability to help students appreciate and celebrate their cultures of origin while gaining knowledge of and fluency in at least one other culture. Finally, sociopolitical or critical consciousness refers to the ability to take learning beyond the confines of the classroom using school knowledge and skills to identify, analyse, and solve real-world problems (Ladson-Billings, 2014). This final domain has particular relevance to the STEM disciplines, alongside addressing the concerns noted in the introduction to this chapter. In parallel with Ladson-Billings, Gay (2002) proposed Culturally Responsive Teaching (CRT), which is defined as using the cultural characteristics, experiences, and perspectives of ethnically diverse students as conduits for teaching them more effectively. According to Pinneo and Benton (2024), CRP and CRT are used interchangeably in this chapter.

Education in the STEM field has historically struggled to integrate elements of multiculturalism into praxis (Charity Hudley & Mallinson, 2017). Similar issues—and responses to them—are also seen in informal or out-of-school educational experiences and more broadly in science communication (Bang et al., 2010; Medin & Bang, 2014). The application of CRP to the STEM field had a higher hurdle than other subjects because "too many teachers and teacher educators think that their subjects (particularly math and science) and cultural diversity are incompatible, or that combining them is too much of a conceptual and substantive stretch for their subjects to maintain disciplinary integrity" (Gay, 2002). However, given that even Western science relies on their cultural and philosophical background (introduced in Chap. 2), clearly the STEM field is not incompatible with CRP. While uncommon, examples of CRP in the science or STEM fields have been reported in five areas: academics, cultural competency, social inequities, culturally relevant learning environment, and rejections of deficit lens (Pinneo & Benton, 2024). For example, Teo and Khoh (2014) proposed culturally relevant science activities using typical foods and materials in Singapore, such as dragon fruit for natural indicators or turmeric spice for natural dye.

In response to the need for an inclusive and comprehensive model, a model of CRP in science and mathematics was developed (Hernandez et al., 2013). The model consisted of five thematic categories: content integration, facilitating knowledge construction, prejudice reduction, social justice, and academic development. For example, Gomes et al. (2024), by applying the model, implemented culturally relevant science activities that included interdisciplinary discussions, experimental activities, and modelling exercises to integrate local knowledge about the use of cactus pears by rural communities in northeastern Brazil.

Finally, Gould et al. (2018) reported the correlation of cultural relevance with environmental behaviour, connection to nature and self-efficacy. This indicated the necessity of further exploring the essentially limitless variants of culture and how those may interact with environmental attitudes and behaviours (Gould et al., 2018). Studies in the Native American context have also productively brought cultural relevance together with learning about the environment, with "the idea that nature is not an externality, apart from humans, but rather that humans are a part of nature" (Bang et al., 2010) emerging as a key design principle—a theme we also explore here. The potential of CRP in environment/sustainability learning is particularly relevant to the case of Kakishibu, given it can be substituted for plastics in many applications.

8.1.2 Research Questions

Two research questions guided this study: RQ1. How effectively can CRP-based practices nurture children's sociopolitical/critical consciousness of global plastic issues and alternatives, and of disaster management? RQ2. To what extent can cultural differences impact this effect?

8.2 Research Methods

8.2.1 Development and Implementation of Culturally Relevant STEM Education Programme on Kakishibu in Australia and Japan

We conducted the study in Japan as well as in Australia to compare the effects of a culturally relevant STEM education programme on Kakishibu for primary school children in different cultural backgrounds. While allowing cross-cultural comparisons in the study, for attendees in Australia, this also fulfilled CRP's second domain of cultural competence, whereby they can appreciate knowledge from other cultures (notwithstanding attendance by many Australians of Japanese descent). Across other domains, sociopolitical/critical consciousness has become the most ignored component (Ladson-Billings, 2021). However, this study focused on sociopolitical/critical consciousness because this is important to enhance children's environmental attitudes and behaviours. Kakishibu is culturally relevant to Japan. On the other hand, Australia is known as a successful multicultural country (Elias et al., 2021). Therefore, the culturally relevant STEM education programme on Kakishibu may positively affect both Japanese and Australian children to varying degrees.

Throughout the workshop, the natural, sustainable, and environmentally friendly properties of Kakishibu and its water repellent properties are contrasted with the fossil-fuel-based alternative of plastics—particularly those used only for a single use then discarded. The country context around plastic use is relevant to reception of the workshop. Regarding plastic consumption per capita, Australia was ranked first, compared with Japan as fifth in 2019 (Charles et al., 2021). Since 2019, every state in Australia has banned single-use plastics such as food and drink containers, cutlery and drink stirrers, food trays, straws, and heavyweight carry bags, often substituting with items made from paper or wood. Japan has introduced charges for plastic bags since 2020 as part of the "strategy for the recycling of plastic resources," developed by various ministries such as the Consumer Affairs Agency, Ministry of Foreign Affairs, Ministry of Finance, Ministry of Education, Culture, Sports, Science and Technology, Ministry of Health, Labour and Welfare, Ministry of Agriculture, Forestry and Fisheries, Ministry of Economy, Trade and Industry, Ministry of Land, Infrastructure, Transport and Tourism, and the Ministry of the Environment in 2019. The wide range of government agencies involved is testament to the complex systems plastic is embedded in. Additionally, under the Act on Promotion of Resource Circulation of Plastics, measures including charging for plastic spoons and other items that were previously handed out free at convenience stores commenced in 2022. Hence, both Australian and Japanese children could find everyday cultural relevance in efforts to reduce plastic use and potential alternatives, but to differing degrees. This difference provided an intriguing extension to the study, whereby enduring traditional knowledge from one culture would be overlayed with everyday cultural practices in the present day.

8.2.1.1 Pre-development of the Programme

Before developing a STEM workshop on Kakishibu based on CRP, we developed and implemented a hands-on workshop on Kakishibu for primary school children (Goto & Kano, 2021). It was a request from Mr. Matsuda, mentioned in Chap. 6. We held the workshop at a shopping mall in Japan in January, 2020. The workshop offered a bouncy ball scooping activity, which is a famous game for children during *Matsuri*, a Japanese festival held at shrines and temples on special occasions. In the game, children scoop bouncy balls floating on the water's surface with a scoop made from plastic with a paper membrane. When the paper breaks under the weight of the balls as it weakens due to the water, the game is over. They try to scoop as many bouncy balls as possible until the paper breaks. At the hands-on workshop, children experienced three hands-on activities: the bouncy ball scooping game with a non-treated paper scoop, dying a paper with Kakishibu and the bouncy ball scooping game with a Kakishibu dyed/treated paper scoop (Fig. 8.1).

The non-treated paper scoop is easily broken (left), while the Kakishibu-treated paper scoop is strong and water-resistant (right).

The workshops were cancelled after the first one, as the COVID-19 pandemic broke out. Instead, we started developing and implementing an online workshop on Kakishibu. We mailed families an experiment kit, including bouncy balls, and paper scoops with and without Kakishibu. We developed a video with subtitles to proceed with the main structure of the workshop so that participants could catch up on instructions even if an online conference tool was delayed due to Internet technical issues. Using the mailed experiment kit and video, undergraduate students at Shiga University facilitated the online workshops for 50 minutes (Kano et al., 2024). The content included a little bit of the history of Kakishibu, but it was primarily a hands-on workshop—while mentioned it was not designed to be overtly culturally relevant.

Fig. 8.1 Comparison between a non-treated paper and a Kakishibu-treated paper scoops

8.2.1.2 Development of the Programme

After the COVID-19 situation has gradually settled both in Japan and Australia, we decided to revise the hands-on workshop and implement it in-person both in Japan and Australia. In revision, we introduced the CRP and made the workshop include Japanese culturally relevant content related to Kakishibu. We also decided to use a video with subtitles to proceed with the main structure of the workshop, as we succeeded in online workshops to make them inclusive (Kano et al., 2024). The video also gave consistency across workshops because we could use the same video spoken in Japanese and English (dubbed) for each country, with English and Japanese subtitles, respectively. This allowed control of the structure and quality of the workshops by reducing the variation in the progress of different workshops. In a methodological sense, this added experimental rigor by (as much as is feasible in real-world workshops) controlling other variables while running the same workshop in the two countries. The video offered eight activities. At each activity, the facilitator (first author of this chapter) facilitated interactive communication and discussion among participants. The workshop was 50 minutes long, including the 20-minute video.

The contents and purposes of the eight activities are shown in the Table 8.1. The videos guiding the eight activities used in the workshops are as follows:

https://www.youtube.com/watch?v=GInpdyO73NE (English dubbed version).
https://www.youtube.com/watch?v=7?j-chXI8ns (Japanese version).

Activities 1, 2, 6, 7, and 8 are designed to be culturally relevant to Kakishibu's history, traditional knowledge, and plastic society. Regarding CRP's three domains (Ladson-Billings, 1995; Ladson-Billings, 2014), activities 1, 2, and 6 are designed to foster children's cultural competencies. Activities 2, 3, 4, 5, 7, and 8 are designed to foster participants' critical consciousness as well as cultural competencies; that is, participants are expected to apply and generate novel ideas to solve real-world problems, such as environmental issues or evacuation in disaster or emergency situations. Moreover, the activities aligned with the CRP model in science and mathematics (Hernandez et al., 2013) by integrating content (all activities), facilitating knowledge construction (all activities), prejudice reduction (mainly in student-student interaction activities 3, 6, 7, 8), social justice (mainly in activity 8), and academic development (mainly in hands-on activities in 1, 2, 4, 5).

We also introduced the utility value intervention technique in the programme to enhance participants' learning motivation. The utility value intervention is "an interactive, classroom-based assignment designed to help students make connections between the content they are learning and their lives" (Hulleman & Harackiewicz, 2021). In striking studies, simply having students write essays connecting their science learning to their everyday lives—with no other pedagogical intervention—resulted in increased interest in science and substantially better grades, especially in students with low expectations of success (Hulleman et al., 2010). Utility value is a dimension of value that motivates students though highlighting the application of learning, either now or in the future (Hulleman & Harackiewicz, 2021). Canning and Harackiewicz (2015) proposed two variations of

Table 8.1 Eight activities and their purposes

No	Activities	Purposes	CRP viewpoints
1	Bouncy ball scooping game with a non-treated paper scoop	Introducing a Japanese culturally relevant game in Matsuri, festival Hands-on	Designed to be culturally relevant to Kakishibu's history, using traditional knowledge To foster children's cultural competencies and connecting to the real-world festival context
2	Bouncy ball scooping game with a Kakishibu-treated paper scoop (note that Kakishibu is introduced as "liquid X" with over 100 years of history)	Introducing a Japanese traditional knowledge of water resistance Hands-on	As above, and also extending the cultural knowledge to frame it as solution to a real-world problem relevant to children (i.e. critical consciousness)
3	Comparing two scoops with and without liquid X (Kakishibu)	Scientific observation using four senses (look, touch, smell and listen) Student-student interactions	Furthering critical consciousness and cultural relevance through comparison and reflection
4	Putting as many balls as possible on the Kakishibu-treated paper scoop	Experimenting with how strong the Kakishibu-treated paper is Hands-on	Connecting the learning activity to everyday life via a challenge/problem-solving scenario (i.e. critical consciousness)
5	Using the Kakishibu-treated paper scoop to prevent the water from splashing out	Experimenting with how water resistant the Kakishibu-treated paper is Hands-on	As above
6	Guess what "liquid X" is made of	Revealing Kakishibu's history and traditional knowledge Student–student interactions	Designed to be culturally relevant to Kakishibu's history, traditional knowledge, and plastic society to foster children's cultural competencies
7	Guess what Kakishibu was used for before plastic took over	Learning more about cultural aspects of Kakishibu usage and comparing old sustainable ways of life with plastic issues in modern society Student–student interactions	Designed to be culturally relevant to Kakishibu's history, traditional knowledge, and contrast current plastic society to foster participants' critical consciousness as well as cultural competencies
8	Come up with ideas for inventions with Kakishibu	Self-generating invention ideas using Kakishibu to address environmental problems or contribute to disaster prevention as good citizens	Designed to be culturally relevant to Kakishibu's history, traditional knowledge, and plastic society to foster participants' critical consciousness as well as cultural competencies

utility value interventions: (1) directly communicating information about everyday utility value and (2) making students self-generate a short essay (up to three paragraphs in length) briefly describing the potential relevance of the learned technique to their own life.

The utility value of a Kakishibu treatment was directly demonstrated in activities 2-5, while in activity 8, we introduced the self-generated utility value intervention. We asked participants to self-generate invention ideas using Kakishibu to solve environmental problems or prevent disasters. This provided a way to investigate the degree of critical consciousness children had at the end of the workshop, i.e. to what degree were children able to take cultural and other knowledge related to Kakishibu and use it to solve real-world problems. We provided a worksheet with examples for participants to draw a picture of their invention and its explanation (Fig. 8.2).

Pictures were selected as the data collection method as they provide a familiar, accessible, open-ended, and creative way for children to communicate their CRP-influenced invention ideas. Being less dependent on language, they are also a valuable approach in cross-cultural studies, particularly compared to surveys or interviews which can blur meaning in translation. A range of studies have used drawings effectively to assess young people's relationship with and aspirations for the environment (Alerby, 2000; Barraza, 1999; Beach & Smith, 2023; Bentz & O'Brien, 2019; Fleer, 2002; Günindi, 2012; Özsoy, 2012; Sorin et al., 2012) and to probe ideas for inventions and problem-solving scenarios (Shaw et al., 2022). Significantly, creating drawings—including during a research process—can build young people's resilience in the face of difficult situations (Mayaba & Wood, 2015), a property we hoped would assist youth as they responded to the challenges of plastic pollution and disasters. The studies above also demonstrate that drawings are an effective method across diverse cultures. Studies focusing on cultural differences including Japanese youth, however, show the cultural environment, nature, and complexity of images in everyday life, and culturally influenced level of drawing experience can result in substantially different drawings between cultures (Wilson & Wilson, 1987).

We collected the worksheets at the end of the workshop with consent from the children's guardians. The worksheets represent participants' sociopolitical and

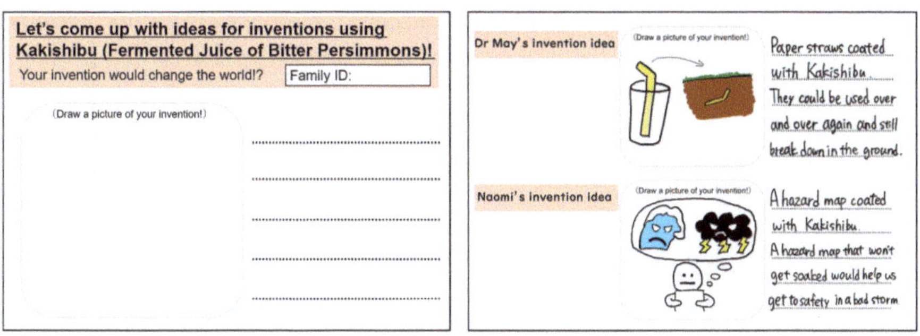

Fig. 8.2 Worksheet for inventions using Kakishibu. Drawing and explanation about inventions using Kakishibu (left) and examples on the reverse side (right)

critical consciousness regarding the connection between traditional knowledge, Kakishibu, and environmental issues or disaster prevention.

8.2.1.3 Implementation of the Programme in Australia

We implemented the developed programme for primary school children (and their families) in Canberra, Australia, during National Science Week 2023 (August 12 to 20). National Science Week is Australia's annual celebration of science and technology, which was established in 1997. It features a vast array of events and experiences, from hands-on activities for families to panel discussions and talks for adults, in-person and online across Australia. We held multiple workshops at four different libraries all over Canberra and at Questacon—The National Science and Technology Centre, which the public could access easily.

8.2.1.4 Implementation of the Programme in Japan

We implemented the developed programme for primary school children mainly in Otsu and Kusatsu cities, and Ryuo town in Shiga prefecture, Japan, from November 4 to 26. In Japan, November is called Cultural Month. Many science and culture events are held throughout Japan every November. We held workshops at the Shiga Museum of Art, the Mizunomori Water Botanical Garden, the Otsu City Science Museum, Ritsumeikan University, and three shopping malls, allowing easy access by the public.

8.2.2 Research Participants

8.2.2.1 Participants in Australia

We held nine individual workshops at five different places across Canberra from August 12 to August 20, 2023. The 86 primary school children were openly recruited through Libraries ACT, the National Science and Technology Centre, and National Science Week 2023 websites (Table 8.2). Participants were required to make reservations through the website. We collected 82 worksheets from 86 children (the collection rate is 95.3%).

8.2.2.2 Participants in Japan

We held 21 programmes at seven different places from November 4 to November 26, 2023. The 357 primary school children were openly recruited through flyers to every primary school in the Otsu and Kusatsu regions, which are two major cities in

Table 8.2 The number of participants in each programme

No.	Number of children	Dates	Places
1	10	August 12	Gungahlin library
2	14		
3	9	August 13	Tuggeranong library
4	12		
5	2	August 15	Gungahlin library
6	5	August 17	Questacon—The National Science and Technology Centre
7	8	August 19	Kippax library
8	12		
9	14	August 20	Belconnen library

Table 8.3 The number of participants in each programme

No.	Number of children	Dates	Places
1	26	November 4	Ritsumeikan University (as one of the events in the science and culture events)
2	18		
3	22		
4	21	November 5	Shiga museum of art
5	20		
6	15		
7	16	November 11	Shopping mall in Ryuo town
8	13		
9	16		
10	20	November 12	Mizunomori water botanical garden
11	13		
12	21		
13	15	November 23	Shopping mall at Otsu city
14	11		
15	14		
16	19	November 25	Shopping mall at Kusatsu city
17	18		
18	19		
19	15	November 26	Otsu City science museum
20	12		
21	13		

Shiga prefecture, and the websites of each place and Shiga University (Table 8.3). Participants were required to make reservations through the website, similar to the procedures implemented for workshops in Australia. We collected 353 worksheets from 357 children (the collection rate is 98.9%). The number of participants in

Japan was higher than in Australia, considering that the Japanese population is about 4.5 times larger than the Australian population.

8.2.3 *Protocol for Qualitative Analysis of Worksheets*

We developed a protocol for qualitatively analysing the collected worksheets. While more exploratory protocols for analysing drawings take a grounded theory approach cataloging all picture elements (e.g. Weber et al., 2011), we instead formed initial codes based on our research questions to assess children's critical consciousness on environmental issues and disaster management. First, we set up two major codes: "Eco-friendly or sustainable perspectives" and "Helping people in case of disaster or emergency" from the top-down viewpoint of our programme's purpose. Second, we co-developed sub-codes from the bottom-up view of the collected worksheets. Very few worksheet drawings focused on "helping people in case of disaster or emergency," hence no sub-codes we required.

Under the major code "eco-friendly or sustainable perspectives," we found four sub-codes reflecting the degree of environmental consideration in the inventions: "Mentioned," "Strongly suggested," "Suggested," and "Not mentioned." The criteria and examples are shown in Table 8.4. We coded one code with each worksheet. When a worksheet includes multiple sub-codes, we assign the highest ordinal level (Mentioned > Strongly suggested > Suggested > Not suggested).

Table 8.4 The sub-codes of the major code "Eco-friendly or sustainable perspectives"

Sub-categories	Criteria
Mentioned	Used when invention/application clearly and explicitly references being eco-friendly and sustainable. This code shows that the innovative and sustainable properties of Kakishibu have been deliberately incorporated into the invention. Examples: Explicitly replacing plastic or other non-sustainable materials for environmental benefits, noting it was good for the environment, explicitly addressing environmental problems or reusability, etc.
Strongly suggested	Used when invention/application suggests being eco-friendly and sustainable, but Kakishibu's sustainable properties are not explicitly and clearly shown/mentioned (e.g. replacing items often made of single-use plastic or other single-use items commonly made of plastic). Examples: Disposable paper products (already used as plastic alternatives), paper alternatives, or other single-use items commonly made of plastic
Suggested	Used when invention/application suggests focusing on plastic products noted as being waterproof and considering the waterproofing property of Kakishibu, but not primarily from eco-friendly perspectives. Examples: Raingears, ocean/pool gears, or other items which may be subjected to water
Not suggested	Used when invention/application suggests considering the waterproofing property of Kakishibu, but from no eco-friendly perspective.

8.2.4 Measure of Agreement

The first author translated the worksheets written in Japanese (ten of eighty-two worksheets from Australia and all from Japan) into English. Two researchers (authors) independently coded all the worksheets according to the protocol, using the MAXQDA 24 software (VERBI GmbH, Berlin, Germany). One coder was Australian and the other Japanese, allowing cultural aspects of drawings to be seen in context. We introduced the weighted Kappa to measure their inter-rater agreement on sub-codes because they are regarded as ordinally scaled.

8.2.5 Research Ethics

The ethical aspects of this research have been approved by the ANU Human Research Ethics Committee (Protocol 2023/388) and Shiga University Human Research Ethics Committee (Protocol B230202).

8.3 Results

8.3.1 Measure of Agreement

We conducted weighted Kappa statistics with sub-codes of "Eco-friendly or sustainable perspectives." We did not conduct the statistics with the "Helping people in case of disaster or emergency" code because of the small number of cases. The result of the inter-rater agreement for worksheets from Australia and Japan is shown in Table 8.5. The results indicated almost perfect or substantial agreement in both cases.

The disagreements were resolved through mutual consultation to determine which subcode to use.

8.3.2 Sub-Codes of "Eco-friendly or Sustainable Perspectives" from Australia and Japan

The results of sub-codes of eco-friendly or sustainable perspectives from Australian and Japanese children's worksheets are shown in Fig. 8.3.

The results of tests of normality to Australian data ($p < 0.01$) and Japanese data ($p < 0.01$) indicated that neither Australian nor Japanese distributions were normal. Therefore, we conducted the Mann-Whitney U Test to compare Australian data with

Table 8.5 The weighted Kappa

Samples	Weighted Kappa	Evaluation[a]
Worksheets from Australian children ($n = 82$)	0.82	Almost perfect
Worksheets from Japanese children ($n = 353$)	0.76	Substantial

[a]We used the following criteria according to Landis and Koch (1977)
0.8≦weighted Kappa≦1: Almost Perfect
0.6≦weighted Kappa<0.8: Substantial
0.4≦weighted Kappa<0.6: Moderate
0.2≦weighted Kappa<0.4: Fair
0≦weighted Kappa<0.2: Slight.
weighted Kappa<0: Poor

Fig. 8.3 The comparison of sub-code distribution between Australia and Japan

Japanese data. The results suggested significant differences between the two ($p < 0.05$). The effect size (r) is 0.12, which indicates a small to medium effect.

The examples of sub-codes mentioned and strongly suggested are shown in Figs. 8.4 and 8.5.

8.3.3 *The Code of "Helping People in Case of Disaster or Emergency"*

We found four cases from Australia and seven from Japan related to the code "helping people in case of disaster or emergency." The examples are shown in Fig. 8.6.

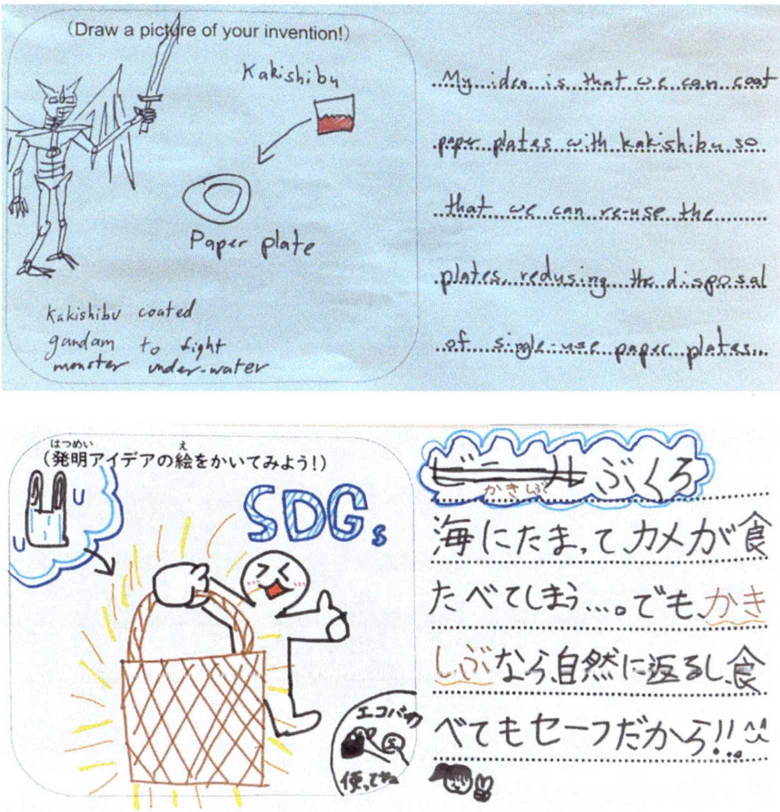

Fig. 8.4 Examples of the sub-code mentioned from Australia (upper) and Japan (lower). Japanese texts mean that "~~Plastic~~ Kakishibu bags: plastic bags accumulate in the sea and are eaten by turtles. On the other hand, Kakishibu bags are safe to eat and return to nature." The striking out of "plastic" then replaced with "Kakishibu" (lower) and diagrammatic treatment of paper to make it re-usable (upper) along with noting of associated environmental benefits indicate a high level of eco-friendliness in the invention. The drawings are published with the participants' permission

8.4 Discussion

This study investigated the application of CRP, principles typically applied to formal learning, to a hands-on, informal STEM learning workshop with sustainability objectives underpinned by cultural and traditional knowledge. Over half of the children from the two countries (67% in Australia and 54% in Japan) came up with ideas suggesting the eco-friendly use of Kakishibu (Fig. 7.3). This indicates that the culturally relevant STEM education programme has positively impacted the majority of children's sociopolitical/critical consciousness. These measures need to be viewed through the lens of being self-generated invention ideas conveyed through

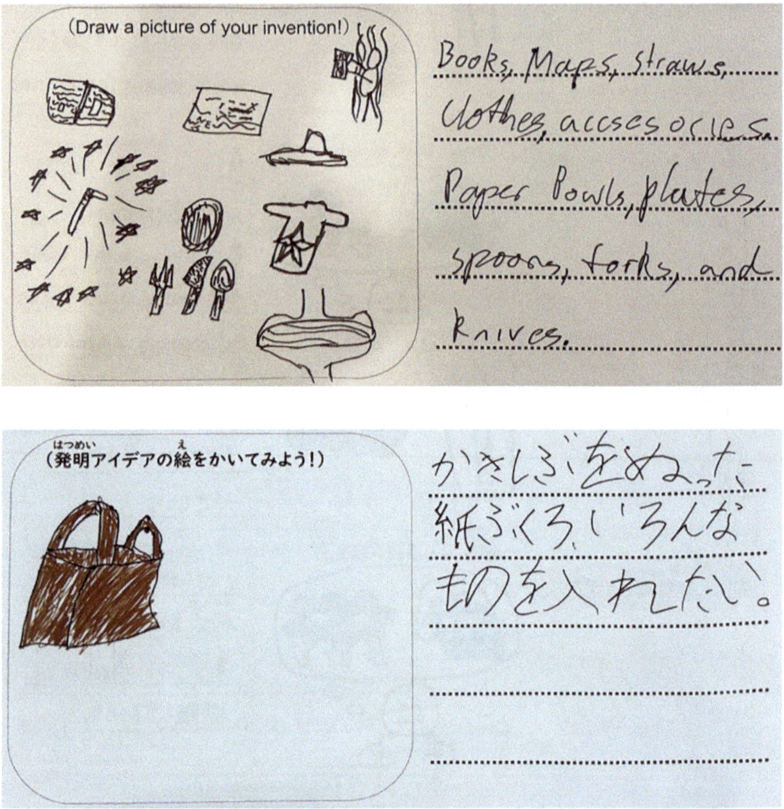

Fig. 8.5 Examples of the sub-code strongly suggested from Australia (upper) and Japan (lower). Japanese texts mean that "paper bags dyed with Kakishibu. I want to put lots of things in the bag." The drawings are published with the participants' permission

the drawing medium. We argue these are a more robust measures of learning and impact than a typical post-event survey with closed questions, rather they represent open-ended, participant-responsive, measures. Open-ended, self-generated relevance (sometimes termed utility and attainment values) is often argued as a more potent or motivational form of relevance particularly in individuals who lack confidence (Canning & Harackiewicz, 2015; Priniski et al., 2018).

We expected the workshop would generate eco-friendly invention ideas. In this respect, the results are encouraging yet unsurprising; however, unpacking findings across the two cultures is quite the opposite. We hypothesised the workshop should have greater effects on Japanese children given the Japanese cultural elements, yet across several coding categories the results showed the converse. Australian children generated more inventions that mentioned (17%) and strongly suggested (28%) eco-friendliness compared to Japanese children (14% and 16%). In contrast,

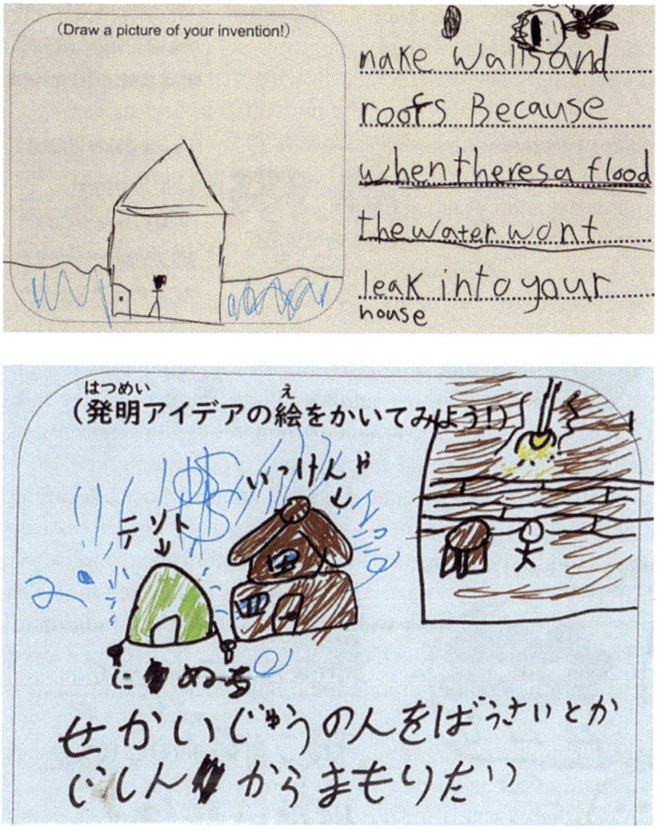

Fig. 8.6 Examples of the code "Helping people in case of disaster or emergency" from Australia (upper) and Japan (lower). Japanese texts mean, "I want to protect people worldwide from disasters and earthquakes." The drawings are published with the participants' permission

Japanese children (46%) generated more inventions that did not suggest eco-friendliness compared to Australian children (33%).

These results suggest that the programme impacted Australian children's socio-political/critical consciousness more strongly, despite the "mismatch" of cultural elements. This is consistent with one of CRP's tenets—that contextualising content in cultures beyond those of the student benefits learning. Moreover, Australia is multicultural, so the CRP-based programme may have been more effective due to incorporating information and examples from different cultures. While inclusive designers of informal learning programmes commonly adapt content to target relevance to different cultures, these results suggest—particularly in mixed or multicultural groups—that simply being more broadly culturally inclusive can yield positive results.

Nevertheless, a deeper explanation can be found in the overlay of contemporary culture in each country. In the strongly suggested code, replacing single-use plastics was an overwhelmingly frequent theme, showing that many children understand the problems with single-use plastics and are motivated to find alternatives. The difference between Australian and Japanese results in the strongly suggested code can largely be explained by inventions related to single-use plastics, which makes sense when looking at the context in each country. As of 2019, Australia had the highest per capita generation of single-use plastic waste anywhere in the world (Charles et al., 2021). A range of policy interventions since 2018 have aimed to ban, recycle, or otherwise reduce plastic waste by 2025—goals that have now been pushed back to 2040 due to poor progress (Anderson & Gbor, 2024). As a result, many areas in Australia have banned common single-use plastics, e.g. shopping bags at supermarkets, and hence, the issue is part of everyday culture, including in the school curriculum. Japan, on the other hand, are ranked fifth in per capita single-use plastic waste generation globally, with less and more targeted policy intervention—hence, the issue is not as front of mind in contemporary culture. So while we expected the traditional knowledge (i.e. Kakishibu)-based elements to be more strongly culturally relevant to children and inspire inventions within their culture of origin, we believe the overlay of contemporary cultural factors has shifted the balance. Hence, science communicators/educators wanting to (re)introduce traditional knowledge into their practice should also consider the nuance of its fit with everyday culture, particularly where attitudinal or motivational outcomes (i.e. critical consciousness) are desired.

A considerable number of children (33% in Australia and 46% in Japan) did not have any suggestions for using Kakishibu in an eco-friendly manner. However, it did not always reflect that they had no ideas for inventions. Instead, many inventions show or imply treating plastic items with Kakishibu rather than replacing the plastic with a Kakishibu-treated cloth or paper. This shows the dominance of the current system and aspects of path-dependency and lock-in in the sociotechnical transition around plastic use (Loorbach et al., 2017). Plastic use is so central in people's thinking they have difficulty readily imagining it being replaced. In other words, plastic is already so dominantly central that it is a somewhat intractable everyday part of culture for children.

On the other hand, some worksheets mentioned the oceans, ocean plastic pollution, microplastics, and similar themes. For a minority of children (17% in Australia, 14% in Japan), the programme promoted a systems approach to viewing the broader issues plastic waste causes. So while the dominance of plastic use can obscure eco-friendly applications of Kakishibu, it can also encourage broader thinking and application to wider environmental problems linked to plastic use.

Finally, a small number of children proposed ideas related to "helping people in case of disaster or emergency." This suggests that most children regarded Kakishibu as relevant to eco-friendliness rather than disaster prevention.

8.5 Implications, Limitations, and Future Research

Our independent coding represented a high quality of agreements (weighted Kappa: 0.82 in Australia and 0.76 in Japan). However, when we examined the disagreements in detail, we realised that this was due to the cultural differences between the two coders. For example, book covers are made of plastic in Australia and paper in Japan, lunch bags are made of plastic in Australia and fabric in Japan, Japanese people more commonly use fabric face masks as well as single-use plastic-based face masks, and eco-bag means reusable plastic bag in Japan even though it is called "eco." This shows the difficulty and limitations of evaluating CRP practice, as well as the importance of collaboration between researchers who are familiar with the culture of each country in which the practice is being carried out.

Replacing single-use plastics was a common theme, showing that many children understand the problems with single-use plastics and are motivated to find alternatives. As these children are the next generation of consumers, this should be noted by manufacturers as an important part of their corporate social responsibility obligations, social license, and business considerations.

Finally, even though eco-friendly was one of the key messages in the workshop, most inventions (except sub-code mentioned) had an anthropocentric approach—children were far more drawn to Kakishibu's properties that could enhance their own or people's lives than helping the environment. This could be due simply to the functional properties communicated in the workshop being better absorbed than the environmental properties. However, it could also suggest a human-nature disconnect where children see themselves as above and not part of nature—Kakishibu is there to serve them first and foremost—which goes against Japanese views of nature that regards humans as part of nature in harmony. As Gienger et al. (2024) pointed out, youth need more than experiences in nature to become more eco-friendly; rather, they need to reorient how they see themselves and humans within nature. Further research and development of CRP-based programmes focusing on Japanese views of nature or Indigenous knowledge, where humans and nature are together in harmony, is needed.

8.6 Conclusions

We developed and implemented a culturally relevant STEM education programme on Kakishibu for primary school children, aligned with the CRP model in science and mathematics. Then, we assessed the practices in Australia and Japan from the viewpoint of sociopolitical/critical consciousness, especially on environmental issues and disaster prevention. We analysed worksheets on invention ideas using Kakishibu with a developed coding protocol.

We proposed two research questions. As for RQ1: How effectively can CRP-based practices nurture children's sociopolitical/critical consciousness of global plastic issues and alternatives? We showed the majority of children suggested the eco-friendly use of Kakishibu. Moreover, some children promoted a systems approach to wider plastic issues.

Finally, regarding RQ2: To what extent can cultural differences impact this effect? Our analysis suggests that the response to CRP-based practices varied largely because of the everyday cultural context around plastic use, rather than the alignment of traditional knowledge and more enduring culture with the cultural identity of participants. The latter was effective for engagement across both countries and a range of cultural identities—cultural relevance provided positive outcomes to all involved, not just the culture incorporated in the STEM content. Our result suggested that the relevance to both multiculture and single-use plastic alternative culture may have led to a synergetic effect.

In sum, these results suggested that the CRP-based STEM programme fostered children's sociopolitical/critical consciousness, while also highlighting how children have been confined within the current sociotechnical situation.

Finally, given children employed an anthropocentric approach rather than focusing on the background philosophy of Kakishibu, Japanese views of nature, we need to emphasise changing the concept of "culturally relevant" from their everyday life to their cultural background philosophy. This would better answer Watanabe's regret for science education in Japan.

References

Alerby, E. (2000). A way of visualising children's and young people's thoughts about the environment: A study of drawings. *Environmental Education Research, 6*(3), 205–222. https://doi.org/10.1080/13504620050076713

Anderson, L., & Gbor, N. (2024). *Plastic waste in Australia*. The Australia Institute.

Bang, M., Medin, D., Washinawatok, K., & Chapman, S. (2010). Innovations in culturally-based science education through partnerships and community. In M. Khine & I. Saleh (Eds.), *New science of learning: Cognition, computers and collaboration in education* (Vol. 1, pp. 1–23). Springer.

Barraza, L. (1999). Children's drawings about the environment. *Environmental Education Research, 5*(1), 49–66. https://doi.org/10.1080/1350462990050103

Beach, R., & Smith, B. (2023). Youth production of digital media to address the climate crisis. *Climate Literacy in Education, 1*(2), 51–57. https://doi.org/10.24926/cle.v1i2.5349

Bentz, J., & O'Brien, K. (2019). ART for CHANGE: Transformative learning and youth empowerment in a changing climate. *Elementa, 7*(1). https://doi.org/10.1525/elementa.390

Canning, E. A., & Harackiewicz, J. M. (2015). Teach it, don't preach it: The differential effects of directly-communicated and self-generated utility-value information. *Motivation Science, 1*(1), 47–71. https://doi.org/10.1037/mot0000015

Charity Hudley, A. H., & Mallinson, C. (2017). "It's worth our time": A model of culturally and linguistically supportive professional development for K–12 STEM educators. *Cultural Studies of Science Education, 12*(3), 637–660. https://doi.org/10.1007/s11422-016-9743-7

Charles, D., Kimman, L., & Saran, N. (2021). *The plastic waste makers index*. Minderoo Foundation.

Elias, A., Mansouri, F., & Sweid, R. (2021). Public attitudes towards multiculturalism and interculturalism in Australia. *International Migration & Integration, 22*, 1063–1084. https://doi.org/10.1007/s12134-020-00784-z

Fleer, M. (2002). Curriculum compartmentalisation?: A futures perspective on environmental education. *Environmental Education Research, 8*(2), 137–154. https://doi.org/10.1080/13504620220128211

Gay, G. (2002). Preparing for culturally responsive teaching. *Journal of Teacher Education, 53*(2), 106–116. https://doi.org/10.1177/0022487102053002003

Gienger, A., Nursey-Bray, M., Rodger, D., Szorenyi, A., Weinstein, P., Hanson-Easey, S., Fordham, D., Lemieux, D., Hill, C., & Yoneyama, S. (2024). Responsible environmental education in the Anthropocene: Understanding and responding to young people's experiences of nature disconnection, eco-anxiety and ontological insecurity. *Environmental Education Research, 30*(9), 1619–1649. https://doi.org/10.1080/13504622.2024.2367022

Gomes, J. P., de Lima, M. S., & Filho, F. F. D. (2024). Integrating local knowledge on cactus pear from Brazilian northeastern communities for culturally responsive chemistry teaching. *Journal of Chemical Education, 101*(8), 3578–3583.

Goto, T., & Kano, K. (2021). Evaluating the representation of participants in a scientific workshop held in a shopping mall: A challenge for attracting a wide range of participants in terms of interests and learning motivation in science. *Japan Journal of Educational Technology, 45*(1), 113–126.

Gould, R. K., Krymkowski, D. H., & Ardoin, N. M. (2018). The importance of culture in predicting environmental behavior in middle school students on Hawaiʻi Island. *PLoS One, 13*(11), e0207087. https://doi.org/10.1371/journal.pone.0207087

Günindi, Y. (2012). Environment in my point of view: Analysis of the perceptions of environment of the children attending to kindergarten through the pictures they draw. *Procedia – Social and Behavioral Sciences, 55*, 594–603. https://doi.org/10.1016/j.sbspro.2012.09.541

Hernandez, C. M., Morales, A. R., & Shroyer, M. G. (2013). The development of a model of culturally responsive science and mathematics teaching. *Cultural Studies of Science Education, 8*(4), 803–820.

Hulleman, C. S., & Harackiewicz, J. M. (2021). The utility value intervention. In G. M. Walton & A. J. Crum (Eds.), *Handbook of wise interventions: How social psychology can help people change* (pp. 116–136). The Guilford Press.

Hulleman, C. S., Godes, O., Hendricks, B. L., & Harackiewicz, J. M. (2010). Enhancing interest and performance with a utility value intervention. *Journal of Educational Psychology, 102*(4), 880–895. https://doi.org/10.1037/a0019506

Kano, K. (2023). Rikabanare saiko: Diversity, equity, inclusion no kanten kara (rethinking of "decline in students' positive attitude toward science": From the viewpoint of diversity, equity and inclusion). *Gendai Shiso (Revue de la pensée d'aujourd'hui), 51*(4), 108–116.

Kano, K., Azai, H., Hagihara, K., Ishikawa, M., Nagayasu, Y., Nishida, M., Tanaka, R., Tsushi, Y., & Yamada, H. (2024). Design and development of an inclusive online STEM course for families. *Research in Integrated STEM Education, 2*(3), 183–205. https://doi.org/10.1163/27726673-bja00026

Ladson-Billings, G. (1994). *The dreamkeepers: Successful teaching for African-American students*. Jossey-Bass.

Ladson-Billings, G. (1995). But that's just good teaching!: The case for culturally relevant pedagogy. *Theory Into Practice, 34*(3), 159–165.

Ladson-Billings, G. (2014). Culturally relevant pedagogy 2.0: A.K.A. the remix. *Harvard Educational Review, 84*(1), 74–84.

Ladson-Billings, G. (2021). I'm here for the hard re-set: Post pandemic pedagogy to preserve our culture. *Equity & Excellence in Education, 54*(1), 68–78. https://doi.org/10.1080/10665684.2020.1863883

Landis, J. R., & Koch, G. G. (1977). The measurement of observer agreement for categorical data. *Biometrics, 33*(1), 159–174. https://doi.org/10.2307/2529310

Loorbach, D., Frantzeskaki, N., & Avelino, F. (2017). Sustainability transitions research: Transforming science and practice for societal change. *Annual Review of Environment and Resources, 42*, 599–626.

Mayaba, N. N., & Wood, L. (2015). Using drawings and collages as data generation methods with children: Definitely not child's play. *International Journal of Qualitative Methods, 14*(5), 1–10. https://doi.org/10.1177/1609406915621407

Medin, D. L., & Bang, M. (2014). The cultural side of science communication. *Proceedings of the National Academy of Sciences, 111*(Suppl. 4), 13621–13626. https://doi.org/10.1073/pnas.1317510111

Naganuma, S. (2015). A study of research trends in "decline in students' positive attitude toward science": Focusing on its current conditions and causes. *Journal of Science Education in Japan, 39*(2), 114–123.

Özsoy, S. (2012). Investigating elementary school students' perceptions about environment through their drawings. *Educational Sciences: Theory and Practice, 12*(2), 1132–1139.

Pinneo, L., & Benton, A. L. (2024). Culturally responsive elementary science teaching: A meta-analysis of current science teaching studies and implications. *Cultural Studies of Science Education, 19*(4), 553–572. https://doi.org/10.1007/s11422-024-10222-6

Priniski, S. J., Hecht, C. A., & Harackiewicz, J. M. (2018). Making learning personally meaningful: A new framework for relevance research. *The Journal of Experimental Education, 86*(1), 11–29. https://doi.org/10.1080/00220973.2017.1380589

Shaw, C., Bernardi, F., & Nickpour, F. (2022). Child-centred framing through design research: A framework for analysing children's 'dream wheelchair' designs to elicit meaning and elevate their voice. *Disability and Rehabilitation: Assistive Technology, 17*(5), 1–13. https://doi.org/10.1080/17483107.2022.2071487

Sorin, R., Brooks, T., & Haring, U. (2012). Exploring children's environmental understandings through the arts. *Creative Approaches to Research, 5*(1), 15–30.

Teo, T. W., & Khoh, R. L. (2014). *Teaching science in culturally relevant ways: Ideas from Singapore teachers*. World Scientific Publishing.

Watanabe, M. (1991). *The Japanese and Western science* (O. T. Benfey, Trans.; E. O. Reischauer, Contributor). University of Pennsylvania Press. (Original work published 1976 as *Nihonjin to kindai kagaku: Seiyo e no taio to kadai* (The Japanese and Western science: Response and challenges to the West), Iwanami Shoten).

Weber, N., Duncan, D., Dyehouse, M., Strobel, J., & Diefes-Dux, H. A. (2011). The development of a systematic coding system for elementary students' drawings of engineers. *Journal of Pre-College Engineering Education Research (J-PEER), 1*(1). https://doi.org/10.7771/2157-9288.1030

Wilson, B., & Wilson, M. (1987). Pictorial composition and narrative structure: Themes and the creation of meaning in the drawings of Egyptian and Japanese children. *Visual Arts Research, 13*(2), 10–21.

Kei Kano, PhD in life sciences, is a professor at the Science Communication Laboratory in the Faculty of Education at Shiga University, Japan. He is a visiting fellow at the Australian National Centre for the Public Awareness of Science at the Australian National University. He also serves as the chair of foreign affairs at the Young Academy of Japan, the Science Council of Japan, and an associate member of the Science Council of Japan. He has won the Prize for Science and Technology from the Ministry of Education, Culture, Sports, Science and Technology (MEXT) in Japan twice, in 2014 and 2017, for his work in science communication practices and research.

Graham Walker is a science communication teacher, researcher, and practitioner based at the Australian National Centre for the Public Awareness of Science (CPAS) at the Australian National University. He works at the interface of science communication and informal STEM learning, focussing on capacity building, emotion and motivation, and communication of social and environmental issues—particularly in informal learning contexts such as science shows and hands-on workshops. He also teaches in these areas. ORCID: 0000-0001-7462-8261, Twitter/Facebook: @ DrGrahams.

Part V
Futures of Green Innovation

Chapter 9
Elements of a Green Innovation Framework: A Vision for Kakishibu in the Plastic Society and Beyond

Sujatha Raman, Bangle Wu (Bernice), and Kei Kano

Abstract This chapter summarises key insights from the main contributions to this volume, putting them in conversation with social science themes related to sustainable transformation and innovation pathways. It outlines a framework for green innovation which brings together social science concepts of system transformation for sustainable futures with concepts informed by the work of the Kakishibu community. This framework underlines the need to simultaneously consider and integrate modes of knowing nature, modes of working to transform nature, and modes of living with nature within innovation paradigms. Such a three-way approach is essential for overcoming the limits of innovation pathways centred primarily on technological breakthroughs without due consideration of their environmental and social burdens. This technology-centred approach threatens to perpetuate patterns of resource over-use to the exclusion of tackling industrial and societal cultures of consumption.

For Kakishibu, a green innovation framework centred on modes of knowing, working, and living highlights the limits of focusing solely on ways of modifying a traditional material to improve its properties. An innovation pathway that only prioritises the aspiration to reproduce the qualities of petrochemical plastic is a dead-end in this regard. A responsible approach to innovating with Kakishibu instead requires the transformation of cultural expectations in industry, research, and society towards a future in which technological ingenuity is devoted equally to maintenance, repair and, overall, to making do.

S. Raman (✉) · B. Wu (Bernice)
The Australian National Centre for the Public Awareness of Science, Australian National University, Canberra, ACT, Australia
e-mail: Sujatha.Raman@anu.edu.au; Bernice.Wu@anu.edu.au

K. Kano
The Australian National Centre for the Public Awareness of Science, Australian National University, Canberra, ACT, Australia

Faculty of Education, Shiga University, Otsu, Shiga, Japan
e-mail: kkano@edu.shiga-u.ac.jp; Kei.Kano@anu.edu.au

K. Kano et al. (eds.), *Traditional Knowledge for Green Innovation with Japanese Persimmon Tannin, Kakishibu*, https://doi.org/10.1007/978-981-95-7805-4_9

Keywords Innovation pathways · Green innovation · System transformation · Sustainability transformation · Technology and culture · Maintenance and repair · Plastics crisis · Kakishibu · Ways of knowing

9.1 Introduction

Petrochemical plastics retain a stranglehold over global production and the use of plastic materials. These plastics have come into circulation in massive quantities over a minuscule 50+ year period in human history, making their toxic presence felt on multiple forms of planetary life, human included. The precision they afford is valuable in cases where high safety standards are important, most obviously in healthcare settings. But the vast majority of plastics comes into existence for far more mundane purposes. Curbing future plastic production is hence a major imperative for activists and experts advocating for a Global Plastics Treaty (e.g., Landrigan et al., 2025). Achieving this goal remains a tall order, with UN negotiations breaking down with weary predictability over the goal of reducing production.

Clearly, we have a long way to go to tackle the planetary crisis of plastics, which demands action across multiple levels from governance, economic incentives, and user norms, to the task of imagining and bringing into being alternative ways of meeting human needs that are no longer wedded to a versatile but hazardous material. This book was motivated by the last of these imperatives, the imperative of green innovation (Leach, 2015).

Green innovation as we use the term in this volume calls for the application of expertise and technological ingenuity in directions necessary for building a future beyond the plastic society. At the outset, we recognise, following the work of archaeologists, that given their toxicity and resistance to breakdown, traces of our petrochemical plastics will endure into deep time (Schofield et al., 2025). In that sense, there is no going 'beyond' plastic society in a timescale discernible to humans and human lifespans. Indeed, some anthropologists and STS scholars investigating the science and management of plastic pollution argue that the challenge is not just one of cleaning up litter—or matter that is 'out of place'. Rather, we need to recognise and grapple with the unexpected entanglements already emerging between organisms ('life') and plastics ('non-life'), some of which are symbiotic (de Wolff, 2017). Similarly, plastics may be designated as 'waste' in one context but become essential sources of livelihood for people in the informal economy in other parts of the world (Holt & Littlewood, 2017).

However, if we are to curtail further indiscriminate production, use and disposal of these plastics, we must also attend to prospects for alternative materials, which could deliver some of the qualities sought out by humans but in a more sustainable manner. As Max Liboiron reminds us in light of an entire pre-Bakelite history of artisanal materials, "plastics have been otherwise" (Liboiron, 2021, p. 3). It is this impulse which underpinned our effort to feature the history of—and contemporary developments around—one such material long used in Japan for purposes similar to those for which we look to petrochemical plastics today. Rather than working from the premise that this material, Kakishibu, can and should replace modern plastics everywhere—a green technological fix (Sarewitz and Nelson, 2008; Rodner, 2004), as it were—chapters in the book featured the voices of a diverse community of Kakishibu experts who seek to revive traditional knowledge of this material and put it to appropriate use in a world transformed by modernity.

In this concluding chapter, we bring together the threads of a possible future for Kakishibu as described and imagined by the visionaries featured in the volume. Woven together, these represent the foundations of a way of thinking about green innovation which strives to overcome simplistic, albeit dominant, narratives of environmental solutions to planetary crises emerging from the mass scale-up of science-led technological breakthroughs achieved in the laboratory.

As we saw in Chap. 7, the industrial laboratory does indeed play a key role in Japan's Kakishibu community. However, the Maxell laboratory project on Kakishibu is notable for its effort to re-imagine the cultural ethos of research and development (R&D) and innovation management. The authors describe their experiment with ways of fusing modern scientific and commercial goals, on the one hand, with traditional techniques and cosmological understandings of human action which have persisted through the centuries, on the other. The significance of *Wakon-Yosai*, the Japanese philosophy of adopting Western technology while preserving culture, comes to the fore in Chap. 2. While this demonstrates the possibility of reconciling tradition with modern science, traditional ideas of humans as part of nature—as opposed to being separate from it—can all too easily lead to a disavowal of responsibility for environmental degradation. The author grasps this challenge by the mettle, arguing that Japan has a particular responsibility to show a way out of global plastic society, given that the country is one of the world's largest producers of plastic waste.

Other chapters extend this idea of cultural work being as critical to the green transformation of societies and innovation ecosystems as scientific knowledge. Chap. 6 considers one such cultural aspect, focusing on inquiry-based learning in schools and student-led collaborations with businesses and local governments to experiment with options for reviving Kakishibu, while Chap. 8 reports on science communication activities with young people and their families in this regard. Chaps. 3, 4, and 5 walk us through the history of Kakishibu and its properties as understood in scientific terms. They cover its traditional mode of craft production and diverse uses in everyday life in Japan prior to the emergence of the modern plastic society in the 1950s, which displaced the use of Kakishibu in most applications. Finally and most importantly, Chap. 5 offers a vivid glimpse of a green innovation future for Kakishibu characterised as much by a transformation of scientific, cultural and societal expectations of how we live and work with nature, as it is by novel possibilities for utilising the properties of this traditional material to meet human needs.

In the rest of this chapter, we briefly outline the elements of a framework for green innovation in conversation with key conceptual insights from the main chapters on Kakishibu and social science debates on planetary crises and innovation. We then distil the implications for future innovation pathways for Kakishibu. We begin in the next section with some background on social science debates on innovation pathways and the role of culture and imagination in this regard.

9.2 Background to a Framework for Green Innovation

The sketch that follows in Sect. 9.3 of a green innovation framework is motivated by two key ingredients. The first is the compelling story of Kakishibu and contemporary efforts, as featured in this volume, to revive its potential. The second is the ongoing debate in the social sciences about the directionality of innovation and the need to actively consider and create space for diverse innovation pathways (Beck et al., 2021; Schot & Steinmueller, 2018; Leach, 2015; Stirling, 2011). These debates build on decades of research in science and technology studies (STS) and the role of innovation in environmental transitions or sustainability transformations. This research highlights the importance of thinking about transformative innovation in socio-technical ways. In this section, we provide an overview of this line of inquiry.

In previous work, one of us described the socio-technical approach to transformation as being attentive to "options for innovating in social, industrial, economic, policy, and cultural systems as much as and in conjunction with technological innovation" (Raman, 2015), p.118. This sentiment was expressed with reference to debates about the ethics and governance of technological breakthroughs in the field of engineering biology (or synthetic biology as it is also known). In these contexts, emerging technologies are put forward as novel options for breaking out of or solving contemporary dilemmas of development and growth on a degrading planet. Yet, they could equally be seen as deeply entrenched in existing industrial systems and hence, *insufficiently* novel for the scale of the challenges we face. From a system innovation lens, a socio-technical perspective opens up space for thinking differently about the forms of novelty we need to cope with planetary crises (Raman, 2015).

The concept of sociotechnical imaginaries offers a way into this challenge. Research in this tradition stresses the foundational role of culture and the imagination in making dominant forms of innovation possible while excluding others (Beck et al., 2021; Jasanoff & Simmet, 2021). In Jasanoff and Simmet's (2021, italics in original) words, "cultures and technological systems *cannot* be understood *apart* from one another. (p.3)" Importantly, this work shows that for dominant material systems to be transformed, we ought to be curious about the tacit imaginaries which hold these systems together. These include sensibilities about what makes a shared sustainable future, a good life and a good society. Petrochemical plastics are tangible forms of an imagined world in which disposability, convenience and consumer choice are (implicitly or explicitly) deemed essential for human flourishing (and hence to the good life) and where the costs and unequal distributional impacts of these forms (insofar as they are considered) are assumed to be optimisable. So to transform this system, we cannot only rely on technological substitution with novel bioplastics which, as we saw in Chap. 1, bring their own environmental and resource-use dilemmas. Putting culture at the centre of transformation involves re-imagining collectively held understandings of how we *know*, how we *work,* and how we *live* in conjunction with the materials and technologies we build and use to reflect these cultural sensibilities.

In distilling the elements of a green innovation framework from the main chapters, we extend sociotechnical ideas developed in relation to emerging technologies to consider essential elements of innovation that begin with already existing forms of technological ingenuity, including those pre-existing the modern industrial age. As we have seen across previous chapters of this book, the Japanese tradition of Kakishibu represents one such ingenious practice. Importantly, in terms of socio-technical imaginaries research, Kakishibu reflects a cultural as well as a material practice and so, any attempt to imagine future pathways will need to consider both of these aspects.

9.3 Elements of a Green Innovation Framework

What is striking from the chapters featuring the voices of key members of the Japanese Kakishibu community is how many of them are already thinking in an integrative way about innovation. In this collaboration, we, as social scientists and science communicators, did not need or try to impose any theoretical sense of how innovation ought to tackle culture and technology simultaneously. Rather, we found that this is how Kakishibu experts describe how things are and how they imagine they should be. Each of the following three elements distilled from their voices reflects cultural beliefs and norms, as well as processes for working with and transforming material things for human purposes.

In this case, if we were to define the challenge for innovation with Kakishibu solely in material terms, it might be as follows: modify Kakishibu so as to extend the life of its functional properties or its performance as an agent for improving the durability and water-resistant qualities of paint or paper (for example). From the perspective of breakthrough technological solutions to planetary crises, this goal might conceivably go even further, calling for an effort to transform fermented persimmon juice altogether by adding synthetic polymers which could potentially allow Kakishibu to reproduce the water-proofness and durability of petro-based plastics. In the process, we might end up with a novel entity in the sense in which planetary-boundaries scientists talk about chemical pollution by plastics and other materials (Persson et al., 2022).

But as we see especially in Chaps. 5, 6, and 7, the authors conceive the problem for innovation differently. They define the challenge in ways that encompass the potential for traditional knowledge and techniques to be recognised and valued in the life and work of citizens, private companies and schools. This in turn places limits on just how far to go in terms of modifying Kakishibu before it becomes a material which no longer resembles the original or captures their impulse to value and work with nature. This sense of ‘enoughness’ in innovation reflects a desire to recognise and integrate the two meanings of *Shizen* (nature) that we learned about in Chap. 2. Importantly, there is a desire to create something novel here, but not in the sense of just releasing a new material entity into the world with unknown consequences.

Rather, the form of novelty being envisaged to address the plastics crisis involves dedicated effort to improve Kakishibu applications but in a way that preserves some critical aspects of traditional cultural expertise and heritage while simultaneously attempting to work with and transform, where possible, parts of modern culture—e.g., commercial practices, societal expectations, and collective consciousness writ large. Put differently, it is about transforming—or trying to transform—social, economic, and cultural systems in conjunction with material systems (Raman, 2015), especially in resource-intensive industrial societies including Japan. By contrast, in rural and peri-urban resource-poor settings such as those explored by Jasanoff and Simmet (2021) in Senegal and India, the cultures that need to be transformed may be those of the agents seeking to drive change from 'above' rather than cultural norms operating on the ground. But in either case, the larger point for green innovation is that one must attend to cultural forms including forms of value and valuation together with material forms.

We now unpack these issues in relation to the three interdependent elements of green innovation in our framework: ways of knowing, ways of working, and ways of living. We don't claim that the innovation efforts featured in the main chapters constitute a move towards a complete and internally consistent system transformation of the plastic age in support of green objectives. That would be impossible, not least for the reasons we have already identified; the lifetime of petrochemical plastics is sufficiently long to have attracted the attention of archaeologists (Schofield et al., 2025). Rather, we are interested in how innovation and novelty are being imagined and practiced in an environment where traditional technologies face significant hurdles from dominant epistemic cultures, social norms, and forms of economic rationality.

9.3.1 Ways of Knowing

Knowledge is, of course, central to innovation. It is critical not only to the development of new technologies, but also to the recovery, maintenance, modification and use of technical and material practices. Chaps. 3, 4, and 5 use modern 'Western' scientific terminologies and concepts to describe the properties of Kakishibu and the array of uses to which it has been put over centuries in Japan. We now know that the tannin produced by ageing and fermentation of bitter/astringent persimmon juice is a polyphenol. We also know that the process of fermentation is assisted by nature, or more specifically, that part of nature known as the microbial world.

But how we know and see nature is shaped by culture. There is no universally true way to understand nature, and therefore, how we think about transforming nature with technology is equally culture-bound. Chap. 2 brought this question to the fore, disentangling the cultural underpinnings of how Japan has historically engaged with Western science and technology, and the lessons for how we ought to approach a future for Japan's own ancient technological prowess as embodied by Kakishibu. This historical account offers the principles of dialogue and integration between different culturally embedded ways of knowing as a central pillar of green innovation.

Chap. 2 tells us that nature as a concept signalling a category separate from humans did not exist in Japan. Until the nineteenth century, the Chinese-derived word in use, *Shizen*, connoted a state of being with no distinction between 'humans' here and 'nature' over there to be looked at and transformed. The encounter with Western science changed this, leaving *Shizen* today with a dual meaning reflecting, in some contexts, the traditional sense of being as is, and in others (notably scientific work), the sense of nature out there as an object of human intervention. Still, as STS scholar Sharon Traweek (1996) reminds us, Japanese scientists did not need recourse to a language and mode of thinking embodied by the definite and indefinite articles of English in order to do physics. Traweek argues it was not necessary to have a generic category of natural objects out there to be investigated—e.g., 'the' cat or 'a' cat—in order to achieve scientific excellence. It is possible to do science without necessarily being wedded to a specific cultural view of nature.

This accommodation of multiple ways of seeing nature in the course of doing science extends also to the Japanese way of adopting Western technology and transforming modern technical know-how in their own image. The concept of *Wakon-Yosai* described in Chap. 2 reflects a particular cultural accomplishment, namely, importing new technologies without importing their Western cultural trappings. Or, in other words, of keeping metaphysics separate from the physical sciences and their material embodiments.

What then are the implications of these cultural differences in ways of knowing and engaging nature for formulating appropriate interventions to address planetary crises via green innovation? Keeping ancient culture and modern technology separate has enabled Japan's technological advancement and global leadership, but at a cost. Petrochemical plastics have been avidly taken up in Japan as elsewhere in the world, contributing to a planetary crisis. Culture has not, so far, been able to provide a corrective to this course of development. Indeed, Chap. 2 points out that far from countering unbridled growth in plastics, the monistic view of humans and nature being in harmony may paradoxically inhibit the emergence of a sense of human responsibility for nature.

In the Western context by contrast, we have a parallel challenge arising from the dualist cultural separation of nature and the human, as opposed to the Japanese monist view. On the one hand, this dualism has enabled human exceptionalism as reflected in ideas and practices of controlling nature for human ends, with these carrying through into green innovation paradigms focused on technological breakthroughs. But on the other, separation also lies at the root of the limitations of parts of the modern environmental movement, insofar as the focus of responsibility is on 'humans' as a singular group 'saving the planet' out there. This too leads to a kind of exceptionalism and attempts to purify nature as separate from culture, generating its own paradoxes.

For example, in her fieldwork with environmental scientists and marine clean-up movements, de Wolff (2017) wonders what will happen to the convict surgeonfish, which have been newly liberated from their plastic habitat. She cautions that a simple clean-up of the environment can generate its own thoughtless consequences if the act overlooks how plastics and organisms have commingled to create new ecosystems of co-dependence or new forms of life. Similarly, plastic waste has become a source of income for some global South communities who make a living by improvising and 'making do' (Holt & Littlewood, 2017). So we can agree that plastic pollution is a problem and a crisis and that its toxic burdens fall disproportionately on the vulnerable, thus creating particular challenges for environmental justice (Abrahms-Kavunenko, 2023). But equally, there is no escaping the need to consider how humans, nature and technology have—and are—co-evolved in tackling even apparently straightforward green objectives such as cleaning up environmental pollution.

These complexities suggest the necessity to cultivate attention to how we produce and bring knowledge to bear on goals of green innovation in light of ambiguous boundaries between nature and culture. Chap. 2 argues that, in the Japanese context, responsibility for transforming the plastics crisis calls for the ability to *differentiate*, when required, between nature and 'artificial nature' or nature that has been transformed through human intervention. Likewise, we might add that in Western contexts, responsibility requires attending to the *entanglements* already created between the same nature and its artificial incarnation and what these mean for future innovation designs. In these cases, the novelty that is needed encapsulates *how* we know as well as *what* we know in designing green innovation.

9.3.2 Ways of Working

Ways of knowing nature fundamentally connect with ways of working with nature in different settings, from manufacturing (Chap. 4) to education and learning (Chap. 6) to research and development (Chap. 7). If ways of knowing constitute a conceptual and culturally embedded (or implicit and taken-for-granted) scaffolding for innovation, ways of working represent how people seek to innovate in practice. How are problems framed for innovation objectives and how do these frames shape the solutions pursued in innovation projects? Are different ways of knowing and making (artefacts, materials, technologies, economic devices) brought to bear on these endeavours and, if so, how? For example, how is the innovation process structured and managed to allow novel ideas to emerge while preserving what ought to be cherished? These questions reflect what we are calling 'ways of working' in innovation.

To explore them, let us first consider the way things currently work in the sphere of interest. At first sight, it may seem that established practices are, almost by definition, *not* a part of 'innovation' as they are precisely what must be modified, improved upon or transformed. Preservation seems antithetical to innovation. But an extensive body of research highlights the myriad invisible and visible ways in which some well-established and seemingly routine practices—notably, those relating to everyday maintenance and repair—are absolutely critical to the performance of modern socio-technical systems. Technologies break down, malfunction, or perform sub-optimally, and it takes work to keep them running. This is as true for modern technologies as it is for pre-modern ones including Kakishibu (as we saw in the previous section). Some scholars point to repair practices as a way of challenging the focus on innovation (e.g., Denis, 2019; Russell & Vinsel, 2018), while others weave together maintenance, repair, and innovation as part of a whole (e.g., Jackson, 2014). Those who work to keep current technological systems going have precious know-how and expertise, the loss of which would be a major blow to future systems.

Bringing the work of maintenance into the picture helps restore the missing elements in how modern technologies are conventionally represented and, consequently, what it takes to innovate in order to tackle the problems they have created. The search for novelty in innovation does not—or should not—mean the erasure of everything that is familiar including expertise honed over many years, or passed down through the generations over centuries.

In our case, we can agree that petrochemical plastics can no longer be represented as disposable, time-saving devices—at least, not without inviting critique from those having to maintain the waste management systems in which they end up and from those peering into the oceans further down the chain or casting light on planetary health in which human health is fundamentally interconnected with nature or the condition of the planet. Similarly, bio-based plastic solutions (innovation) to the problem of petro-plastics raise their own issues for waste management depending on how well these bio-based materials break down, how long they take to do so, under what conditions and with whose labour. Seen from this lens, we can bring a more system-wide perspective to understanding and valuing traditional techniques (such as for production, modification and use of Kakishibu) and putting them in dialogue with efforts to imagine and formulate innovation pathways. Drawing from the chapters, we highlight two critical aspects of ways of working in this manner towards green innovation.

First, rather than frame the problem for innovation as finding a way to radically break with the past, it might instead be framed as about finding a way to fuse traditional techniques, modern science and commercial application (Chap. 7). As we saw in Chap. 5, there are ways that producers and users traditionally worked with Kakishibu to extend its lifetime in different contexts by adding other natural materials. However, the dilemma for innovators today as explored in Chap. 7 is that these natural materials 'fall short' by the standards of synthetic fibres made from petrochemicals. Yet, such standards do not represent universal human needs abstracted from context as Jasanoff and Simmet (2021) observe in a different but related case of sustainable energy transitions. The high functionality and performance of modern plastics is only possible with the use of synthetic chemicals, with all the attendant environmental and social problems which are outsourced to other parts of the system.

So the challenge as defined by the engineers at Maxell (Chap. 7) and the high school students and teacher they collaborated with (Chap. 6) was how to prevent the loss of cultural know-how around the production of Kakishibu, increase the economic value of a traditional industry in modern commercial settings, stay as close as possible to the goal of minimising the use of synthetics, and communicate the value of their invention in ways that would resonate with consumers. Early on, they recognised that while alternatives to petrochemical plastics are emerging on the market, these bio-derived materials are still synthetic polymers with unknown ecosystem impacts (see also Chap. 1 on this point). However, they faced distinct challenges making a case for a different approach based on Kakishibu but with the intent to stick to natural/biological materials for modification or innovation pathways.

The commercial environment in which they worked meant that research and development of novel Kakishibu-based materials would need to fit within the company's main area of business. At first sight, the fit was not obvious given that Maxell specialises in electronics and electrical devices, while current applications of Kakishibu have tended to be in food, agriculture and paper. Nonetheless, the team were able to identify Maxell's patented expertise with coating materials as the relevant link, using kraft paper as a promising candidate for incorporating traditional Japanese know-how and extending the company's business opportunities. With this as their focus, they were supported to investigate novel ways of incorporating oils extracted from plants, in order to improve the water-repellent qualities of Kakishibu when used to treat paper.

Their effort exemplifies a second novelty in the way of working towards green innovation, namely, one that extended to how the creative process was itself imagined and managed. Adding a drying agent such as perilla oil can easily make the paper product prohibitively expensive, an unattractive proposition for commercial scale production when much cheaper synthetic materials are readily available. How then to create an opening for Kakishibu while keeping costs down? Chap. 7 highlights the path to an innovative workable solution, with a dual layer of coatings being the first step followed by application of emulsifying techniques, which meant they could dispense with the need for expensive drying agents and rely instead on more ordinary oils. How though did the team stumble upon the ideas informing their experiments and eventually leading to a successful patent application? Dr Goto cites the influence of Buddhist philosophy as mediated through the prolific writings of Minakata Kumagusu (1867–1941), who is belatedly being discovered as a naturalist scholar who crossed conventional boundaries well before this came to be encouraged under contemporary labels such as 'transdisciplinarity' or 'art-science' engagement (Honda, 2019).

In Minakata's cosmological view (based on the Buddhist concept and image of the mandala), novel discoveries emerge from chance encounters which are themselves made possible by some sort of universal "force" (as Dr Goto calls it in Chap.

7) or "tact" (as characterised in anthropological STS work on Minakata: Jensen et al., 2016). There is a "heightened attentiveness" (Jensen's words) at play in the process of creation or discovery, with the person's perceptions and ability to grasp the nature of things being modified in the course of the encounter. Other people might encounter the same things without necessarily sparking the same receptivity or, in turn, discovery. For Dr Goto, this meant managing the process of research and development in a way that allowed junior interns to have the freedom to inhabit this sense of an openness to the world, to really notice a phenomenon (in this case, of how camellia leaves repel rainwater on contact) and to allow oneself to be transformed by it (the 'aha' moment).

Similarly, the very idea for the experiment conducted in Maxell's laboratory emerged from an encounter between Dr Goto and an economics professor who introduced him to Mr Matsuda (author of Chap. 5), a Kakishibu aficionado working with his high school students to raise awareness of this traditional material and its properties and to spark ideas for contemporary applications.

Based on these observations, we want to highlight that attending to 'ways of working' in the innovation process helps researchers stand back from taken-for-granted framings of problems that invite innovative solutions in response to planetary crises. This does *not* mean we simply escape those current frames. Rather, the possibility of real systemic innovation, however incomplete, arises from staying grounded in core values (in this case, a commitment to minimising the use of synthetics), being open to diverse perspectives (reflected here by collaboration between students, teachers, interns and industry engineers) and to experiencing the world, including its familiar dimensions, anew (being inspired by a traditional material, Kakishibu, or 'seeing' the ordinary event of rainwater falling on a leaf as if for the first time, all of which allowed a novel thing to emerge).

Being attentive to ways of working also illuminates the need for innovative thinking and design to be brought to bear in other parts of the innovation system outside the laboratory, for example, in production processes. Chap. 5 underlines the case for labour-saving measures in the challenging task of picking persimmon fruit before they ripen (as Kakishibu relies on astringent qualities found in the fruit in its raw/green state). In addition to the height of the trees, fruit-picking imposes further hazards on workers as harvesting happens in peak summer which means coping with high heat. The author therefore suggests the appropriate design and use of technologies such as greenhouse and hydroponic cultivation, robot harvesters and fruit thinning techniques. Similarly, conditions in factories can also be hazardous due to the acidic nature and strong odours of long-life fermentation, again needing the design of appropriate solutions.

In sum, these observations about different workplaces in the Kakishibu system—from the orchard to the factory to the industrial laboratory—point towards a vision of novelty as represented by the fusion of the modern and the traditional rather than a simple flip from one to the other, again reflecting a sophistication that we don't usually encounter in technology-hype.

9.3.3 Ways of Living

Chap. 5 is devoted to ideas for new frontiers for Kakishibu, covering the entire value chain. This includes improved options for growing and picking astringent persimmon; ways to improve the properties of the fermented juice for applications in the global tannin industry; revitalising the Kakishibu industry; creating a space for stakeholders to share knowledge and expertise; and activities to raise wider awareness and increase consumption. But there is a distinctive cultural objective holding these different aspirations together, distinctive in that it is not typically found in mainstream innovation manifestos or imaginaries. This is the author Professor Matsuo's invitation to reflect on traditional Kakishibu cultures as a way of stimulating societal change towards "slow living".

In the innovation paradigm outlined in Chap. 5, Kakishibu is valued not only for its material properties and opportunities for utilising them as a plastic substitute, but also for what it can teach us about forms of life or what we are calling 'ways of living'. In this way, Matsuo's vision is far more novel than is the case with more conventional innovation narratives centred on technological breakthroughs. Green transformation involves working with Kakishibu, exploring suitable options to fuse the tannin with other natural materials and "building *a new natural material culture*" which contributes to "replacing the currently dominant plastic culture" (Chap. 5, italics added).

This desire to fold ways of living into the innovation vision is motivated by a sense that modern societies are founded on the promise of convenience and comfort, which many uses of petrochemical plastics exemplify. Furthermore, we might add that comforts for some come at the expense of hardship for others. Nor is this state attributable simply to individuals. Insofar as innovators imagine human needs to be immutable, they reinforce and reproduce resource-intensive ways of living by supplying 'solutions' which make it possible to ignore temporally and spatially distant impacts. Sociologists emphasise that changing resource-intensive patterns is not merely a matter of changing individual attitudes or behaviour. As sustainability scholar Elizabeth Shove (2010) has argued, patterns of consumption are co-constituted by changing infrastructures of provision and social practices, which become reproduced over time. Through this co-evolution, particular norms of what is taken to be necessary for a good life become stabilised. Over the lifetime of petrochemical plastics, manufacturers have created new uses for plastic, which have in turn been normalised through patterns of consumption.

But if plastic consumption norms have been sustained and reproduced through time and through co-evolution with infrastructures of supply and governance, how then can cultural metamorphosis be possible? There are two dimensions to this challenge of imagining and doing green transformation differently. The first is to tackle the problem of green innovation paradigms, which remain fixated on technological substitutes to the exclusion of their own unequally distributed burdens on nature and society. The second is to be willing to try and bring about a change of collective consciousness in conjunction with material change, as Professor Matsuo has articulated. Transformation in this imagination is system-wide as it incorporates both everyday practices and socio-technical infrastructures, demand as well as supply (Shove & Walker, 2010).

Professor Matsuo expresses the hope that reviving Kakishibu can help reclaim the mindset of that time in Japan's history with "nature as a being to coexist alongside of" (Chap. 5). In terms of the concepts we encountered in Chap. 2, this brings to mind the goal of fusing Western and traditional Japanese ideas of nature. System transformation is founded on cultural change in the sense of a shift in philosophical stance on nature–human relations. But this is not merely an abstract position, involving as it does equal attention to a variety of practical matters including how to: "promote demand for Kakishibu products, design these products, build a mass production system, and then implement price setting and container and packaging design" (Chap. 5).

These expansive system-wide aspirations are motivated not only by the environmental challenges posed by petrochemical plastics but by the recognition that it will be well-nigh impossible to replace them like-for-like with a traditional material like Kakishibu, even in modified form. Kakishibu is "*not* a versatile industrial material" (Chap. 5, italics added) in the way that plastics made from coal or oil are, nor is it as durable. What is novel in this approach is the recognition of natural limits and the desire to tackle a situation in which industries and entire societies have become accustomed to high-performance plastics, and innovators assume the only way through is to focus on transforming nature while leaving culture alone.

The desire to improve within natural limits appears to capture the seemingly limited spirit of 'incremental' as opposed to 'breakthrough' innovation, in the terms famously outlined by the economic historian of technology, Nathan Rosenberg. But this is only the case if we discount the normalising impacts of breakthrough innovations, at least those which are resource- and energy-intensive and which ignore the knock-on effects on communities bearing the accompanying burdens of making technology possible. As Professor Matsuo points out, trying to radically modify Kakishibu with synthetic chemicals will have its own material consequences: the manufacturing process will require excessive energy inputs, and the need to handle chemical waste at the end of the lifecycle has not been solved.

9.4 Future Pathways for Kakishibu in the Plastic Society and Beyond

We are now able to distil the implications of our green innovation framework for future research and innovation pathways for Kakishibu. As we have observed, one could imagine a pathway in which Kakishibu's history of traditional expertise comes to be mobilised within a Western dualist paradigm of a sharp separation of nature and culture. Within this framework of conquest of nature, experimental knowledge may be produced to devise and test ways to improve the properties of

Kakishibu to fulfil human needs. Chap. 5 identifies reasons why this pathway of modifying or improving on traditional Kakishibu mixtures might be attractive at first, but ultimately undesirable.

Kakishibu deteriorates on exposure to sunlight, so there have been various techniques employed through history to stabilise the fermented juice for longer by mixing it with other materials such as straw, wood ash or metal salts. This impulse to tinker and improve things in the material world is an essential part of human evolution. As Professor Matsuo observes (Chap. 5), there are many interesting questions to investigate from an experimental mindset, for example, how quickly Kakishibu-treated products decompose under a variety of conditions. This too represents a way of knowing that has been instrumental in human history. From the standpoint of green innovation today, Chap. 5 highlights some important research questions that are being considered. These focus on ways to enhance and strengthen Kakishibu's properties through mixing with other materials as well as through processes such as bonding or adhesion. For example, "combinations with various proteins (including fibrous proteins), natural oils, polysaccharides (viscous HMW substances), safe minerals, among others, is promising." There are many ideas here for future research opportunities working with Kakishibu and experimenting with ways to improve the traditional material.

At the same time, the desire to improve nature needs to be balanced with ideas of technical ingenuity as 'making do' and staying within limits. The authors of this volume warn against interventions harnessed within the paradigm of conquest of nature. In this context, the goal of innovation isn't so much to find the best possible techniques for modifying Kakishibu to meet the highest imaginable standards of stability and water-proofness of the kind we have been accustomed to in the plastic society. Rather than aiming to extend 'artificial nature' (as Chap. 2 puts it) even more, the innovation goal would be to limit modification of Kakishibu with synthetic chemicals, and scaling up responsibly rather than making excessive demands on Kakishibu (Chap. 2). These demands cannot be met beyond a certain point without completely changing its integrity as a traditional material fused with particular ways of knowing, working and living.

Such a future innovation pathway would be grounded in the recognition that Kakishibu is indeed not a versatile industrial material like petrochemical plastics (Chap. 5), but that the costs of completely overcoming these limits to versatility and durability would be too great. Similarly, a mass production pathway would likely mean the loss of knowledge, expertise and livelihoods as currently embodied in craft production by *Shibuya* or artisans (Chap. 4). A future pathway modelled on responsible innovation might extend also into the laboratory and commercialisation paradigms as we saw in Chap. 7. It will require the valuation and creation of communication and engagement networks (Chaps. 6 and 8) in which students, craft producers, technicians and researchers are brought together to foster awareness of Kakishibu in a way that combines historical legacies, cultural mindsets and scientific and techno-economic aspirations. Chap. 6 outlines a utopian vision for creating a Kakishibu Hamlet in which these values come together and serve as stimulus for opening up novel and better futures.

9.5 Conclusion

Scientific knowledge and technological ingenuity are essential for tackling the plastics crisis. But as we have highlighted in this chapter, a green innovation framework fit for this purpose will need to devise novel pathways to draw from and integrate history, culture, and traditional knowledge as well as modern science. Green innovation in this sense will require a judicious rethinking of societal and epistemic expectations of 'perfecting' nature for human purposes. Equally, thinking this way can in turn open up alternative ways of recognising and valuing what is taken to be novel and desirable.

In her essay for the Toshiba International Foundation 30th anniversary, Eiko Honda (2019) writes stirringly of her vision of 'Knowledge without Supremacy'. She argues that nineteenth-century disciplines are outdated at a time when we are being called upon to put planetary nature at the heart of our ways of knowing. In his posthumously published novel, *The Pale King*, David Foster Wallace wrote: "True heroism is minutes, hours, weeks, year upon year of the quiet, precise, judicious exercise of probity and care – with no one there to see or cheer." Wallace was talking about the humdrum reality of modern life where we are compelled to outgrow the fantasies of childhood stories of the hero's journey. One could equally borrow this sentiment to describe what we are called upon to accomplish in everyday life beyond the confines of modern plastic-reliant society. Kakishibu demands a kind of quiet heroism from us, not just in tolerating the travails of modern society as Wallace would have it, but in transforming our modes of living, working and knowing in sustainable directions.

References

Abrahms-Kavunenko, S. (2023). Toward an anthropology of plastics. *Journal of Material Culture, 28*(1), 3–23.

Beck, S., Jasanoff, S., Stirling, A., & Polzin, C. (2021). The governance of sociotechnical transformations to sustainability. *Current Opinion in Environmental Sustainability, 49*, 143–152.

de Wolff, K. (2017). Plastic naturecultures: Multispecies ethnography and the dangers of separating living from nonliving bodies. *Body & Society, 23*(3), 23–47.

Denis, D. J. (2019). Why do maintenance and repair matter? In *The Routledge companion to actor-network theory* (pp. 283–293). Routledge.

Holt, D., & Littlewood, D. (2017). Waste livelihoods amongst the poor – Through the lens of bricolage. *Business Strategy and the Environment, 26*(2), 253–264.

Honda, E. (2019). Knowledge without supremacy: Japanese studies in the face of global ecological crisis. *Toshiba International Foundation 30th anniversary essay contest on the future research agenda for Japanese studies*. https://cdn2.hubspot.net/hubfs/384373/pdf/activity-report/2019/Honda_Eiko_Essay.pdf

Jackson, S. J. (2014). Rethinking repair. In *Media technologies: Essays on communication, materiality, and society* (pp. 221–239).

Jasanoff, S., & Simmet, H. R. (2021). Renewing the future: Excluded imaginaries in the global energy transition. *Energy Research & Social Science, 80*, 102205.

Jensen, C. B., Ishii, M., & Swift, P. (2016). Attuning to the webs of En: Ontography, Japanese spirit worlds, and the "tact" of Minakata Kumagusu. *HAU: Journal of Ethnographic Theory, 6*(2), 149–172.
Landrigan, P. J., Dunlop, S., Treskova, M., Raps, H., Symeonides, C., Muncke, J., & Rocklöv, J. (2025). The Lancet Countdown on health and plastics. *The Lancet, 406*(10507), 1044–1062.
Leach, M. (2015). What is Green? Transformation imperatives and knowledge politics. In I. Scoones, M. Leach, & P. Newell (Eds.), *The politics of green transformations* (1st ed., pp. 25–38). Routledge.
Liboiron, M. (2021). *Pollution is colonialism*. Duke University Press.
Persson, L., Carney Almroth, B. M., Collins, C. D., Cornell, S., De Wit, C. A., Diamond, M. L., & Hauschild, M. Z. (2022). Outside the safe operating space of the planetary boundary for novel entities. *Environmental Science & Technology, 56*(3), 1510–1521.
Raman, S. (2015). Responsive novelty: Taking innovation seriously in societal research agendas for synthetic biology. *Journal of Responsible Innovation, 2*(1), 117–120.
Rodner, L. (Ed.). (2004). *The technological fix: How people use technology to create and solve social problems*. Routledge.
Russell, A. L., & Vinsel, L. (2018). After innovation, turn to maintenance. *Technology and Culture, 59*(1), 1–25.
Sarewitz, D., & Nelson, R. (2008). Three rules for technological fixes. *Nature, 456*(7224), 871–872.
Schofield, J., Gorman, A., Antonello, A., Couceiro, F., Leterme, S., MacGregor, M., & Werner, A. D. (2025). Archaeological and systemic context for the plastic age: Theorising the formation of contemporary and future archaeological records. *Cambridge Prisms: Plastics, 3*, e33.
Schot, J., & Steinmueller, W. E. (2018). Three frames for innovation policy: R&D, systems of innovation and transformative change. *Research Policy, 47*(9), 1554–1567. https://doi.org/10.1016/j.respol.2018.08.011
Shove, E. (2010). Beyond the ABC: Climate change policy and theories of social change. *Environment and Planning A, 42*(6), 1273–1285.
Shove, E., & Walker, G. (2010). Governing transitions in the sustainability of everyday life. *Research Policy, 39*(4), 471–476.
Stirling, A. (2011). Pluralising progress: From integrative transitions to transformative diversity. *Environmental Innovation and Societal Transitions, 1*(1), 82–88. https://doi.org/10.1016/j.eist.2011.03.005
Traweek, S. (1996). Unity, dyads, triads, quads, and complexity: Cultural choreographies of science. *Social Text, 46/47*, 129–139.

Sujatha Raman is a Professor and Director of the Australian National Centre for the Public Awareness of Science (CPAS), Australian National University (ANU) and UNESCO Chair-holder in Science communication for the Public Good. She is an interdisciplinary social scientist interested in how matters of sustainability and equity intersect, and what these intersections mean for re-imagining how we bring science and innovation to bear on the task of building better planetary futures. She has published widely on responsible innovation and emerging technology assessment, the science/policy interface, and energy transitions and their material underpinnings. Before joining ANU in 2018, she was Co-director of the Institute for Science and Society (ISS), University of Nottingham (UK) and Director of the Leverhulme Research Programme, '*Making Science Public*' which spanned the Universities of Nottingham, Sheffield and Warwick.

Bangle Wu [Bernice] is a PhD candidate at the Australian National Centre for the Public Awareness of Science (CPAS), Australian National University. She has a multidisciplinary background in communication and natural resources management, and previously worked as an intern journalist in the environmental section of Southern Weekly. Her research examines the role of science in environmental regulation, with a focus on per- and polyfluoroalkyl substances (PFAS), widely known as "forever chemicals." Through interviews with scientists and regulatory experts in Australia and internationally, she has investigated how PFAS are problematised through scientific evidence, standards, and interpretations. She has contributed commentary on PFAS regulation in *The Conversation* and the Special Broadcasting Service (SBS). More broadly, she is interested in how alternative ways of knowing might inform and transform our responses to planetary crises.

Kei Kano , PhD in life sciences, is a professor at the Science Communication Laboratory in the Faculty of Education at Shiga University, Japan. He is a visiting fellow at the Australian National Centre for the Public Awareness of Science (CPAS) at the Australian National University. He also serves as the chair of foreign affairs at the Young Academy of Japan, the Science Council of Japan and an associate member of the Science Council of Japan. He has won the Prize for Science and Technology from the Ministry of Education, Culture, Sports, Science and Technology (MEXT) in Japan twice, in 2014 and 2017, for his work in science communication practices and research.

Epilogue

Sujatha Raman

In some ways, the research collaboration from which this volume emerged could be said to have similar origins to the innovation process described in Chapter 7. In this chapter, Dr Goto notes the influence of noted Japanese biologist and humanist, Minakata Kumagusu, on his research management philosophy and describes the opportunities he tried to create to allow new ideas to emerge. His approach paid off handsomely with a successful patent application for a technique to improve the water-repellent qualities of Kakishibu while working with natural materials. A book is not quite the same as a patent, of course. But this volume similarly arose out of the unexpected coming together of people with different perspectives and material associations, from which something new emerged.

Through the chance encounter of a workshop presented by Professor Kano to demonstrate the properties of Kakishibu through hands-on methods, I was moved and inspired to notice a way of connecting with other things I cared about, namely, the possibility of diverse innovation pathways in the world. Likewise, my PhD student at the time, Bernice Wu, was moved by the possibility of finding alternative ways of being in the face of toxic 'forever chemicals', the subject of her thesis. Indeed, I would now say I had only a faint idea of what I meant in 2015 when I imagined system transformation as something that encompasses culture, economy, technology, and so on in a short Perspective article in the *Journal of Responsible Innovation*. I was impelled to say those words at the time, perhaps by some 'force' (in Minakata's terminology via Dr Goto), and the words made their way into print. But it is only upon engaging with the Kakishibu community and re-reading their work, presented here in their own voice, that I can begin to glimpse something about what system transformation could mean.

S. Raman
The Australian National Centre for the Public Awareness of Science,
Australian National University, Canberra, ACT, Australia
e-mail: Sujatha.Raman@anu.edu.au

K. Kano et al. (eds.), *Traditional Knowledge for Green Innovation with Japanese Persimmon Tannin, Kakishibu*, https://doi.org/10.1007/978-981-95-7805-4

Professor Matsuo offers a vision for Kakishibu in which this material comes to be harmonised with a particular cultural mindset that values circular economies and a genuine sustainability—or least, efforts to take sustainability seriously and develop pathways that bring technology and culture together. The sentiments we have featured in conjunction with scientific accounts of Kakishibu—including its production, modification, and communication—chime well with emerging movements in Japan and globally (e.g., the World Sufficiency Lab) which focus on building a future around sufficiency or an ethos of 'enoughness' in both technology development and wider society. We cannot predict how the sufficiency movement will unfold nor how the future of petrochemical plastics will turn out in the coming decades, although we do know that plastics have already transformed the planet and will have an enduring presence into the future. As we look towards curtailing their use, we hope that the story of Kakishibu featured in this volume and the green innovation framework we have developed in conversation with social science research will provide inspiration.

MIX
Papier aus verantwortungsvollen Quellen
Paper from responsible sources
FSC® C105338

If you have any concerns about our products,
you can contact us on
ProductSafety@springernature.com

In case Publisher is established outside the EU,
the EU authorized representative is:
Springer Nature Customer Service Center GmbH
Europaplatz 3, 69115 Heidelberg, Germany

Printed by Libri Plureos GmbH
in Hamburg, Germany